Frederick A. A. Skuse

Diptera of Australia

Part II. - The Sciaridae

Frederick A. A. Skuse

Diptera of Australia
Part II. - The Sciaridae

ISBN/EAN: 9783741183928

Manufactured in Europe, USA, Canada, Australia, Japa

Cover: Foto ©berggeist007 / pixelio.de

Manufactured and distributed by brebook publishing software
(www.brebook.com)

Frederick A. A. Skuse

Diptera of Australia

DIPTERA OF AUSTRALIA.

By Frederick A. A. Skuse.

Part II.—THE SCIARIDÆ.

(Plate xi.)

This contribution, like its predecessor, in no way pretends to be more than an introductory review of the group; but in making immediate use of such material as has been collected, however inadequately that may represent the actual extent of the Australian Sciaridæ, this beginning may at least furnish a basis for future advancement.

The amount of work hitherto done amongst the Sciaridæ of this continent, is evidenced by the record of only a single species, namely *S. reciproca*, Walk., the description of which is absolutely useless, and it is questionable if the name attached to Mr. Walker's type-specimen should be retained, unless the species be re-described.

It is somewhat remarkable that the naturalists of the "Novara" and "Eugenie" expeditions in their collections of Australian Diptera did not obtain here any Sciaridæ, or even examples of the next family, the Mycetophilidæ, more particularly as these two groups elsewhere, and some equally obscure and small flies here, were not completely overlooked.

From the appended descriptions it will be seen that 42 species are enumerated and described as new; if to these must be added Mr. Walker's species, it brings the total up to 43, but this must bear only a small proportion to the unknown number of forms prevalent in the neighbourhood of Sydney alone. The family is no doubt largely represented in Australia, though apparently not so numerously as the Cecidomyidæ, but no peculiar Australian genera have been yet detected ; with the exception of one species which I refer to the genus *Trichosia*, Winn., all belong to the typical genus *Sciara*, Meig.

A large proportion of the species described in the following pages were obtained by Mr. Masters and myself whilst searching for Cecidomyidæ, and therefore the result cannot be regarded as consequent upon very special research ; had the latter been the case, the number would no doubt have been augmented extensively.

My attention has been so completely occupied with the collection and description of the perfect insects that little opportunity has favoured an investigation of the life-histories and young stages of the Sciaridæ; however, I hope the time may not be long postponed when I shall be able to supplement this imperfect work with accounts of these, but to ascertain the complete life-history of individual species is alone the work of months. Meanwhile I have given general descriptions of the larva and pupa stages, entirely summarized from the works of the few authors who have at all studied them, and I hope that the information will tend to direct the attention of students to this most neglected but deeply interesting group, the species of which during the young stages of their existence especially demand our consideration.

In the descriptions I sometimes employ the words "apparently no pubescence," by which I mean that no hair is made visible through the application of an ordinary entomological lens, but is rendered so when submitted to a working microscope of moderate amplification ; indeed but for the latter even the longitudinal rows of pubescence on the thorax of many species could not be made out. I use the term "petiole" to mean that portion of the third longitudinal vein between its origin and the base of its fork ; and "fork" is always an abbreviation for fork of the third longitudinal vein.

CLASSIFICATION OF THE SCIARIDÆ.

The Sciaridæ* form the second family of the Nematocerous Diptera, and at present comprise seven genera, *Sciara*, Meig.; *Trichosia*, Winn.; *Cratyna*, Winn.; *Corynoptera*, Winn.; *Bradysia*, Winn.; *Epidapus*, Hal.; and *Zygoneura*, Meig.; including a large number of species, which in their form and colour vary but little from a common character; indeed, all but a few known species are absorbed by the typical genus *Sciara*.

The genus *Sciara* was founded by Meigen in Illiger's Magazine, (II., p. 263, No. 12). Shortly afterwards Latreille gave this genus the name of *Molobrus*, and although Meigen's name, from its priority, is the rightly accepted one, Westwood, as late as 1840 (Mod. Class. Ins. II.) retains *Molobrus* and discards *Sciara* as a synonym.

During the year 1830, Meigen (Syst. Beschr. VI. Suppl.) described several additional European species of *Sciara*, which he primarily classified according to a very limited estimation of the venation in the wings, while the ultimate separation of the species was entirely restricted to the consideration of coloration. In this imperfect method of dealing with such peculiarly approximate and persistent forms, Meigen was more or less followed by other authors, and as an inevitable consequence, the determination of the species described by these authors, is not only a matter fraught with great difficulty and uncertainty, but in many cases their identification is altogether impossible.

Before 1867, the genera *Sciara*, *Epidapus*, and *Zygoneura* oscillated in an irregular manner between two distinct families, the Cecidomyidæ and the Mycetophilidæ, their affinities being regarded with such uncertainty by Dipterologists. Meigen (Syst. Beschr. I., XXXVI.) placed *Sciara* in a tribe by itself which he designated Tipulariæ lugubri, and he located this tribe between

---

* From σκιαρος, shaded.

43

T. fungicolæ (Mycetophilidæ), and T. latipennes (Simulidæ).
Macquart (Hist. Nat. Ins. Dip. I., p. 121,) classed *Sciara* with
the T. fungicolæ, but *Zygoneura* found a place with the T. galli-
colæ; he nevertheless remarks that *Zygoneura* approaches the T.
fungicolæ by the venation of the wings.   Halliday included all
three genera, together with the genera constituting the sub-family
LESTREMINA of the Cecidomyidæ, amongst the Mycetophilidæ.
Loew, in 1862 (Mon. Dipt. N. America I., p. 13), also puts
Sciara in the Mycetophilidæ, but observes that it differs most
from the rest of the family, and shows some affinity with Cecidomyidæ.   With regard to *Epidapus*, this author says that "it is
quite impossible to place it among the Mycetophilidæ, as Walker
does, if we characterize the families as we have done ; it rather
seems to find its place among the Cecidomyidæ."   *Zygoneura* is
regarded by both Loew and Osten-Sacken as belonging to the
second section of the Cecidomyidæ.   Again, Schiner admits *Sciara*,.
Epidapus and *Zygoneura* into the Mycetophilidæ, the genera of
which he arranges under two sub-families SCIARINA and MYCE-
TOPHILINA.

Winnertz in 1867 (V. z.-b. G., Wien, Band XVIII.) came to
the rescue, and published a monograph of the European Sciaridæ,
in which he described 157 new species, and re-described 30 species
of previous authors.   To this indefatigable Dipterologist we are
indebted for an exhaustive investigation of the three genera which
had hitherto been the source of such perplexity.   Not only did
he establish five new genera and characterize the whole as forming
a distinct family, but he elaborated a system of classification for
the species, and pointed out what characteristics are possessed by
the individual parts of their structure.

The chief divisions of Meigen are split up into sub-divisions
based upon the colours of the halteres and palpi, and into further
sections by the position of the cross-vein, and into sub-sections by

the position of the tip of the second longitudinal and the tip of the lower branch of the fork of the third longitudinal vein, which with rare exceptions are found constant in each species.*

In the ultimate separation of species, characters were found in the length of the costal vein beyond the tip of the second longitudinal vein, as compared with the distance between the tip of the former and that of the anterior branch of the third longitudinal fork ; also that the relative distances between the tip of the posterior branch of the third longitudinal fork and the tip of the anterior branch of the fourth longitudinal fork, and this latter from the tip of its posterior branch, were valuable to notice. Besides the consideration of the alar-vein system, Winnertz also has shown that the relative lengths of joints of the legs are important in specific distinction.

With reference to the two main divisions of *Sciara*, Winnertz adds a note of which the following is a translation :—" In some species the tip of the sub-costal (first longitudinal) vein joins the costa opposite or beyond the root of the fork in the ♀ and in front in the ♂ ; in a few species its position, as well also that of the cross-vein, is not at all constant. In these cases the respective species are mentioned according to their deviation in the corresponding division, but they are described in that division to which they belong according to their majority."

Winnertz (Mon. der Sciarinen, p. 10), after characterising the family Sciaridæ draws attention to the close affinity of this group with the Mycetophilidæ, but at the same time gives the following diagnosis of the characters which supply effective points of distinction, and he justly concludes that these deviations give the Sciaridæ a type so different from that of the Mycetophilidæ, that a combination of the two groups must appear inadmissible.

---

* Winnertz adds, "the position of the tips is determined by drawing a straight line from the middle of the base of the wing through the apex, upon which, as the foundation, perpendicular lines are drawn from the respective points."

1. The position of the head, which is placed less deep in the prothorax.

2. The shape and character of the antennæ.

3. The less high and less acclivous metathorax.

4. The less elongated coxæ.

5. The strongly developed holding forceps of the ♂.

6. The shorter sub-marginal (first longitudinal) vein.

7. The medial vein (the portion of the second longitudinal behind the cross-vein) which is close to the sub-marginal (first longitudinal) vein and nearly always running parallel with it.

8. The fork (of the third longitudinal vein) which is always long-stalked, and generally coming from the middle of the medial vein (the portion of the second longitudinal behind the cross-vein).

As far as the small number of Australian species known to me are concerned, the system employed by Winnertz has been found perfectly serviceable; it only remains to be seen what new divisions fresh forms may require, but it appears to me probable that very little diversity, if any, will be found, judging from the general conformity of those I now describe. However, from my necessarily limited acquaintance at present with the totality of our species, there is little need for me to give expression to any views on the distribution of the group.

CHARACTERS OF THE FAMILY.

THE TRANSFORMATIONS.

I. *Larva.*

The larva is slender, cylindrical, smooth and shining, more or less translucent, white, pale yellow or citron-coloured, and composed of 13 segments. The head is considered the first segment, the three following segments represent the thorax, and the nine remaining constitute the abdomen. Stigmata scarcely visible, arranged a little above the lateral line, one pair on the first thoracic segment and a pair on each segment of the abdomen except the two last. Pseudopodia on the thoracic division and the last abdominal segment; the abdominal segments are also provided with minute pads which are evidently to assist progression. Head black, small, retractile, furnished with a labrum, a pair of transverse dentate mandibles, maxillæ, a somewhat indistinct labium and rudimentary antennæ. Perris, in his description of the larval form of *S. convergens*, notices "a round areola above the insertion of the mandibles which appeared to be the seat of a completely invisible antenna." In giving an account of the larva of *S. Bigoti*, Laboulbène remarks that when disturbed the grubs move themselves with vivacity; their bodies becoming viscous when they are seized. They also have the power of stiffening and straightening themselves. As far as I have been able to ascertain comparatively little has been done towards a knowledge of the young stages of the Sciaridæ, and but few of their life-histories have been completely worked out. The study of the mouth-parts is regarded as a very difficult one, and there is a considerable amount of uncertainty, amongst capable judges, concerning other organs, as is evinced by their discordant opinions.

As a general rule the larvæ are gregarious, and their food is of a vegetable character, though there are instances of their being found in dung. Laboulbène found the larvæ above referred to in a flower-pot filled with ordinary manure which was peopled already with the larvæ of *Aphodius fimetarius*, and he tells us that they prefer that part where the manure is most moist. Perris found *S. convergens* under bark, amidst the excrements and detritus left by the larvæ of a species of *Tomicus*, of which it made its food ; and the same author bred two other species, one from a twig previously inhabited by the larvæ of *Tomicus ramulorum* and *Anobium longicorne*, and the second from a decaying stump full of the dejections and detritus of other larvæ. A large number of larvæ mentioned by Winnertz in his monograph of the family were obtained from under the bark of trees, in decaying vegetable matter, old wood ; others were found in manure

and fungi. Westwood (Mod. Class. Ins. II.) says that he has
observed the "transformations of several species of *Molobrus*,
Latr. (*Sciara*, Meig.), the larvæ and pupæ of which are found
under the bark of felled trees, or at the root of decayed veget-
ables." Olivier (Prem. Mém. sur quelques Ins. qui attaquent les
Céréales, 1813) bred three species of *Sciara* from wheat.
Macquart (Hist. Nat. des Insectes Dipt. I.) says that the larvæ
of *Sciara* develop in the earth ; this is confirmed by Schiner (In
Beitr. Mon. der Sciarinen von Joh. Winnertz), when he declares
that the garden soil is seldom free from them. To undergo their
metamorphosis into the pupa state some of the larvæ construct a
cocoon, but others do not ; of four species mentioned by Bouché
only one makes a cocoon. Dufour (Ann. des Sc. Nat. 2nd Ser.
t. 12, 1839), in a paper on the metamorphoses of Diptera, describes
the stages of *S. ingenua*, the larva of which constructs a cocoon.
According to Perris, *S. convergens* envelopes itself in a whitish,
pellucid cocoon, which it makes in the detritus. The cocoon is
not formed of filaments, but of a mucous substance which the
larvæ secrete, after the manner of those of *Sciophila*.

II. *Pupa.*

The pupa is naked, oblong, and exhibits a general appearance of
the different parts of the imago,—the eyes, antennæ, rudimentary
wings and the feet being plainly distinguishable. The pupa is at
first yellowish-white, amber-yellow, orange or pale reddish ; afterwards the above-mentioned organs become brown. Two more or
less distinct horns appear near the base of the antennæ in most
pupæ. Surface of the abdomen minutely shagreened, with microscopic asperities, the last segment bifid. Stigmata generally
indistinct. The thorax splits for the whole of its length on the
emergence of the imago. The pupæ have a close resemblance to
those of the Cecidomyidæ.

III. *Habits and habitats of the perfect insects.*

The perfect insects are obtainable in Australia all the year round,
but the greatest number of species and individuals have been
obtained in the neighbourhood of Sydney during the spring of the

year. They seem to chiefly haunt shady situations, and have
been frequently found in caves associated with Cecidomyidæ,
Tipulidæ, etc., but never to my knowledge have they been seen to
voluntarily attach themselves to cob-webs in the manner observed
amongst their near relatives, the gall-gnats ; on the contrary, I have
frequently taken both struggling and dead specimens from webs.
A large number of Sciaridæ may be found flying about underneath
dense bushes, on logs, round tree-trunks, and amongst grass, but
as their habitats are for the most part umbrageous, these small,
often minute, sombre-coloured flies are rendered difficult of detection,
and consequently their collection is scarcely an easy matter. An
inspection of windows generally rewards the collector of Diptera
with an abundance of small flies, especially if the windows over-
look a garden or rural expanse ; I have in this way obtained in
one afternoon specimens of a score or more species, the Cecidomyidæ and Sciaridæ being chiefly represented. Insects which
could otherwise be followed only with a remote chance of success,
even by one possessing remarkably acute eye-sight, are readily
seen on a window, and their capture is easily accomplished. The
flight of the Sciaridæ greatly resembles that of the Cecidomyidæ,
and it is often impossible to distinguish between the two, more
particularly if the individual be small.

With regard to the geographical distribution of Sciaridæ, I
might mention that examples of the genus *Sciara* have been
recorded from all the great continents, and many islands more
or less remote from the mainland. They appear to be generally
diffused over the earth's surface, occurring in arctic, temperate
and tropical regions. About thirty species have been named from
North America, and almost as many from South America; others
are known from Africa, South Asia and the Eastern Islands, and
Professor Hutton has described one species from New Zealand.
According to Van der Wulp, *Sciara thomæ*, Linn., a European
species, occurs also in Sumatra. Of the other genera only one,
Trichosia, as far as I know, has been detected out of Europe,
with a single species from North America, and another described
by me in the present paper, but, this far from demonstrates a
restricted range.

IV. *Imago.*

External structure.

Head small, above broader than long, round when viewed from the front, narrower than the thorax. Hypostoma and front broad. Eyes reniform, broader below than above, approaching on the front or contiguous. Ocelli three, arranged in a triangle on the vertex, the lower one smaller than the upper two. Proboscis short, thick, usually slightly projecting; large suctorial labella. Palpi prominent, incurved, four-jointed; first joint very small : second and third joints nearly of equal length, the former generally narrowed at the base, the latter sub-cylindrical or elliptical; the fourth joint slender, elongate; more or less densely covered with most microscopic pubescence, generally sparingly setose, apparently never glabrous. Antennæ arcuated, projecting forward, generally short, 2- + 14-jointed; the joints of the scapus almost bare, prominent, the first joint cyathiform, cupuliform or sub-cylindrical, the second cya-thiform; the flagellar joints cylindrical or ovate, densely pubescent, often verticillate-setose; generally sub-sessile, the pedicels rarely very distinct, sometimes sessile. Thorax ovate, gibbose, with two or three longitudinal rows of setaceous-hairs, more or less interspersed with short fine hairs; the lateral margins between the origin of the wings and the humeri generally with long setaceous hairs, also a few on the scutellum; no transverse suture; scutellum small. Halteres * large, with microscopic pubescence, usually very sparingly setose; altogether wanting in *Epidapus.* Legs long, frequently very long, slender. Coxæ somewhat elongate, except in *Epidapus,* with a more or less sparse seta-ceous pubescence in front. Femora moderately robust, with a shallow furrow on the inner side, covered with microscopic pubescence, setose in front. Tibiæ and tarsi very densely covered

---

*Winnertz says, "Schwinger unbedeckt," but this is not the case even though we may not recognize their often dense microscopic pubescence.

with microscopic pubescence ; the former with or without lateral
spines, and having terminal spurs, the latter furnished with weakly
developed ungues, the pulvilli being small, scarcely perceptible, or
altogether wanting. Wings longer or shorter than the abdomen,
incumbent, generally rounded at the base and apex, but sometimes
cuneiformly narrowed at the former ; microscopically haired,
rarely distinctly pubescent ; ciliated round the margin ; pellucid,
more or less deeply tinted with different shades of brown, and
occasionally hyaline ; generally beautifully iridescent ; altogether
wanting in *Epidapus*. Costal vein never quite reaching the
apex of the wing, its termination distinct. The number of longitudinal veins amounts to four, though often a rudimentary fifth
is more or less distinctly perceptible immediately behind the
fourth ; the third and fourth longitudinal veins furcate. First
and second longitudinal veins and costal vein very distinct. First
longitudinal vein short, joining the anterior margin either before,
at, or a little beyond the middle of the costal vein, or before, over,
or somewhat beyond the base of the fork of the third longitudinal
vein. Cross-vein usually distinct, situated either before, at, or
beyond the middle of the first longitudinal vein. Second longitudinal vein always terminating at some point in the margin
before the tip of the costal vein, seldom forming a fork near the
tip by sending out a short anterior branch into the costa ; that
portion before the cross-vein nearly always running parallel to the
first longitudinal. Third longitudinal vein usually originating
about midway between the base of the second longitudinal and the
cross-vein ; generally pale and more or less indistinct ; the petiole
always long, often longer than the anterior branch of the fork ;
branches of the fork inclined posteriorly, more or less undulated,
and both reaching the margin below the tip of the costal vein, the
tip of the anterior branch being sometimes at, but never before,
the apex of the wing ; the base of the fork more or less cunei-
form, rarely bulbous. Fourth longitudinal vein generally pale,
branching near the base. A distinct longitudinal wing-fold lies
between the fourth and rudimentary fifth longitudinal veins, much
nearer the former. Abdomen composed of seven segments,

clothed with a short pubescence; in the ♂ almost cylindrical, more or less dilated towards the middle, with strongly developed holding-forceps; in the ♀ acuminate, the ovipositor provided with small terminal lamellæ.

The Sciaridæ generally present a uniform livery of some shade of brown or black, and in the venation of the wings strikingly remind us of the Cecidomyidæ belonging to the sub-family LESTREMINA, and of the Mycetophilidæ in their generally microscopically pubescent membrane. The largest known Australian example measures nearly five lines in expanse, and the smallest rather more than a line.

The following synopsis is appended to set forth Winnertz's distribution of the genera :—

A. Flagellar joints of the antennæ cylindrical, pedicelled, or sessile.

SCIARA, Meig.—Wings longer than the abdomen, their surface microscopically pubescent; wing-lobes more or less developed. Joints of the antennæ pubescent.

TRICHOSIA, Winn.—Wings as in *Sciara*, but their surface distinctly hairy.

CRATYNA, Winn.—Wings as in *Sciara*, but the cubitus (second longitudinal vein) united with the costa by a radial vein.

CORYNOPTERA, Winn.—Wings claviform, their surface microscopically pubescent; antennæ of the ♂ pedicelled, verticillate.

BRADYSIA, Winn.—Wings narrow, shorter than the abdomen, their surface microscopically pubescent.

EPIDAPUS, Hal.—Wings and halteres wanting.

B. Flagellar joints of the antennæ in the ♂ ovate, with long pedicels, in the ♀ cylindrical, sessile.

ZYGONEURA, Meig.—Wings as in *Sciara*, but the large fork bellied at the base and its branches undulated

The following is a tabulation of the Australian species of *Sciara*.

I. FIRST LONGITUDINAL VEIN JOINING THE COSTA OPPOSITE OR BEYOND THE BASE OF THE FORK.

(Nos. 96 to 99).

A. Halteres black or brown, the stalk wholly or partly yellow, yellowish or whitish.

(Nos. 96 to 98).

1. Palpi black or brown.

(Nos. 96 to 98).

A. Cross-vein situated before the middle of the first longitudinal vein.

a. *Tip of the second longitudinal vein nearer the apex of the wing than the tip of the posterior branch of the fork.*

(No 96).

C. Cross-vein situated beyond the middle of the first longitudinal vein.

c. *Tip of the posterior branch of the fork nearer the apex of the wing than the tip of the second longitudinal vein.*

(Nos. 97 and 98).

B. Halteres yellow or whitish.

(No. 99).

II. FIRST LONGITUDINAL VEIN JOINING THE COSTA BEFORE THE BASE OF THE FORK.

(Nos. 100 to 136).

A. Halteres black or brown, the stalk wholly or partly yellow, yellowish or whitish.

(Nos. 100 to 134).

1. Palpi black or brown.

(Nos. 100 to 122).

B. Cross-vein situated at the middle of the first longitudinal vein

b. *Tip of the second longitudinal vein and tip of the posterior branch of the fork equally near the apex of the wing.*

(No. 100).

C. Cross-vein situated beyond the middle of the first longitudinal vein.

a. *Tip of the second longitudinal vein nearer the apex of the wing than the tip of the posterior branch of the fork.*

(Nos. 101 to 104).

† *Thorax with two longitudinal rows of hairs.*

(Nos. 101 and 102).

†† *Thorax with three longitudinal rows of hairs.*

(Nos. 103 and 104).

b. *Tip of the second longitudinal vein and tip of the posterior branch of the fork equally near the apex of the wing.*

(Nos. 105 and 106).

c. *Tip of the posterior branch of the fork nearer the apex of the wing than the tip of the second longitudinal vein.*

(Nos. 107 to 122).

† *Thorax with two longitudinal rows of hairs.*

(Nos. 107 to 110).

†† *Thorax with three longitudinal rows of hairs.*

(Nos. 111 to 122).

2. Palpi yellow.

(Nos. 123 to 134).

B. Cross-vein situated at the middle of the first longitudinal vein.

b. *Tip of the second longitudinal vein and tip of the posterior branch of the fork equally near the apex of the wing.*

(No. 123).

c. *Tip of the posterior branch of the fork nearer the apex of the wing than the tip of the second longitudinal vein.*

(No. 124).

C. Cross-vein situated beyond the middle of the first longitudinal vein.

a. *Tip of the second longitudinal vein nearer the apex of the wing than the tip of the posterior branch of the fork.*

(No. 125).

b. *Tip of the second longitudinal vein and tip of the posterior branch of the fork equally near the apex of the wing.*

(Nos. 126 to 130).

c. *Tip of the posterior branch of the fork nearer the apex of the wing than the tip of the second longitudinal vein.*

(Nos. 131 to 134).

B. Halteres yellow or whitish.

(Nos. 135 and 136).

1. Palpi black or brown.

(No. 135).

B. Cross-vein situated at the middle of the first longitudinal vein.

c. *Tip of the posterior branch of the fork nearer the apex of the wing than the tip of the second longitudinal vein.*

(No. 135).

2. Palpi yellow.

(No. 136).

C. Cross-vein situated beyond the middle of the first longitudinal vein.

b. *Tip of the second longitudinal vein and tip of the posterior branch of the fork equally near the apex of the wing.*

(No. 136).

Genus 1. SCIARA, Meigen.

Sciara, Meigen, Illiger's Magazine II., 263 (1803); *Molobrus,* Latreille, N. Dict. d'H. n. (1804); *Sciara,* Macquart, Hist. n. des Ins. Dipt. I., p. 147 (1834); Zetterstedt, Dipt. Scand.; Walker, I., B.; Schiner, F.A. Dipt., 1864; Winnertz, V. z.-b. G., Wien, Band xviii., p. 11.

Head small, roundish, front somewhat flattened; hypostoma and front broad. Eyes reniform, broader below than above, approaching on the front, or contiguous. Ocelli three, arranged in a triangle on the vertex, the lower one smaller than the two upper ones. Proboscis slightly projecting. Palpi short, prominent, four-jointed, the first joint very small, second and third joints almost of equal lengths, the last joint more or less elongate. Antennæ arcuated, projecting forward, 2- + 14-jointed, always longer in the ♂ than in the ♀; the joints of the scapus cyathiform, almost bare, those of the flagellum cylindrical, pubescent, sessile or sub-sessile, the last joint elliptical or elongate. Thorax ovate, gibbose; scutellum small. Halteres large, with a microscopic pubescence, usually very sparingly setose. Abdomen seven-segmented, in the ♂ almost cylindrical or more coniform, with holding-forceps; in the ♀ acuminate; ovipositor generally long, with terminal lamellæ. Legs slender, frequently very long. Coxæ somewhat elongate. Femora with a shallow furrow on the inner side. Tibiæ with small spurs, with or without lateral spines. Last joint of the tarsi with pulvilli. Wings large, microscopically hairy, rounded at the base and apex. Fork of the third longitudinal vein more or less cuneiformly narrowed towards the base. Fifth longitudinal vein imperfect or altogether wanting.

I. FIRST LONGITUDINAL VEIN JOINING THE COSTA OPPOSITE OR
BEYOND THE BASE OF THE FORK.

A. Halteres black or brown, the stalk wholly or partly yellow,
yellowish or whitish.

1. Palpi black or brown.

A. Cross-vein situated before the middle of the first longitudinal vein.

a. *Tip of the second longitudinal vein nearer the apex of the
wing than the tip of the posterior branch of the fork.*

96. SCIARA MACLEAYI, sp.n. (Pl. XI., fig. 1).

♂.—Length of antennæ...... 0·075 inch   ... 1·89 millimètres.
Expanse of wings........ 0·180 × 0·065 ... 4·56 × 1·66
Size of body............... 0·150 × 0·030 ... 3·81 × 0·76

♀.—Length of antennæ...... 0·050 inch   ... 1·27 millimètres.
Expanse of wings........ 0·180 × 0·065 ... 4·56 × 1·66
Size of body............... 0·200 × 0·035 ... 5·08 × 0·88

♂.—Antennæ deep reddish-brown, with a short yellowish
pubescence; rather slender, half the length of the body; basal
joints with a very sparse pubescence; flagellar joints sub-sessile,
2 to 3 times as long as broad, densely pubescent. Head black,
sub-nitidous. Eyes almost contiguous above. Palpi deep brown.
Thorax black, sub-nitidous, with three almost parallel rows of brown
hairs reaching nearly to the scutellum, also some long hairs along
the lateral margins from the humeri; humeri indistinctly tipped
with reddish-brown; pleuræ deep umber-brown; scutellum rather
densely covered with long hairs. Halteres entirely umber-brown
sprinkled with short hairs. Abdomen black or very deep brown,
densely clothed with a moderately long brown pubescence; as
broad as the thorax; forceps wider than the abdomen, umber-
brown densely covered with a short pubescence. Legs deep umber-
brown, all the joints densely pubescent. In the fore-legs the tarsi
somewhat longer than the tibiæ; in the intermediate and hind-
legs the tibiæ a little longer than the tarsi. Spurs shorter than
the fourth tarsal joint. First joint of the tarsi 2½ times the

length of the second ; second joint about $\frac{1}{4}$ longer than the third and shorter than the fourth and fifth together; fourth joint consideralby shorter than the fifth. Wings pellucid, with an almost fuliginous tint, the costal and two first longitudinal veins deep brown ; brilliant roseous and smaragdine reflections when viewed at a certain obliquity. First longitudinal reaching the costa a short distance beyond the base of the fork and opposite to the tip of the anterior branch of the fourth longitudinal vein ; petiole paler than the fork and longer than the anterior branch ; posterior branch short and nearly straight ; tips scarcely divergent. fy* the same length or almost imperceptibly shorter than gh; kl rather more than $\frac{2}{3}$ the length of lm.

♀.—Antennæ short, rather slender, not as long as the head and thorax combined ; flagellar joints sub-sessile, very little longer than broad, but towards the end about $\frac{1}{4}$ longer than broad; terminal joint twice as long as broad. Abdomen obscure castaneous ; lamellæ of the ovipositor deep brown, elliptical. fy slightly longer than gh; kl about $\frac{2}{3}$ the length of lm. The remainder as in the ♂.

Hab.—Lawson and Glenbrook, Blue Mountains, also Bowral (Masters); Manly, near Sydney (Skuse).

C. Cross-vein situated beyond the middle of the first longitudinal vein.

c. *Tip of the posterior branch of the fork nearer the apex of the wing than the tip of the second longitudinal vein.*

97. Sciara sedula, sp.n. (Pl. XI., fig. 2).

♀.—			
Length of antennæ......	0·050 inch	...	1·27 millimètres.
Expanse of wings.........	0·130 × 0·050	...	3·30 × 1·27
Size of body............ ...	0·120 × 0·020	...	3·04 × 0·50

---

* The letters *f, g, h, k, l, m* have been adopted by Winnertz to indicate the tips of the second longitudinal, the costal, the anterior and posterior branches of the fork of the third longitudinal, and those of the fourth longitudinal respectively. This will be made clear by reference to the diagram on plate XI.

Antennæ deep brown, with a short pale yellowish pubescence; slender, rather longer than the head and thorax combined; joints of the scapus very sparsely haired; flagellar joints sub-sessile, twice as long as broad, the terminal joint 3 times longer than broad. Head black, levigate. Eyes contiguous above. Palpi deep reddish-brown. Thorax black, levigate, with two longitudinal rows of brown hairs reaching almost to the scutellum, also some long hairs between the humeri and the base of the wings; humeri very slightly tipped with deep reddish-brown; scutellum black or deep brown, levigate, with long hairs. Halteres brown, sparsely haired. Abdomen very deep reddish-brown, sometimes appearing almost black. Coxæ black or very deep brown, with long brown hairs on the front. Femora, tibiæ and tarsi very obscure pitch-brown, densely haired. In the fore-legs the tibiæ and tarsi of almost equal length; in the intermediate and hind-legs the tibiæ a little longer than the tarsi. Spurs honey-yellow, about the same length as the fourth tarsal joint. First joint of the tarsi 3 times the length of the second; second joint ½ longer than the third; third and fifth joints of equal length and ⅓ longer than the fourth. Wings pellucid with a greyish-brown tint, somewhat pointed at the apex; brilliant smaragdine, rosy, and golden reflections when viewed at a certain obliquity. Costal and two first longitudinal veins almost cinereous. First longitudinal vein reaching the costa immediately opposite the base of the fork; cross-vein very distinct; petiole thicker and less distinct than the fork, shorter than the posterior branch; both branches thicker and less distinct at their base, running almost parallel to one another, slightly divergent at their tips, the posterior branch less arcuated at the base than the anterior one. *fg* three times the length of *gh;* *kl* somewhat shorter than *lm.*

Hab.—Gosford (Skuse). February.

98. Sciara sororia, sp.n.

♀.—Length of antennæ......	0·040 inch	...	1·01 millimètres,
Expanse of wings.........	0·095 × 0·040	...	2·39 × 1·01
Size of body...............	0·080 × 0·015	...	2·02 × 0·38

44

Antennæ deep brown, with short yellowish pubescence; slender, half the length of the body; joints of the scapus very sparsely haired; flagellar joints sub-sessile, very little longer than broad at the base, towards the tip twice as long as broad, the terminal joint longer. Head black. Eyes contiguous above. Palpi black, or very deep brown. Thorax black, levigate, with two rather indistinct rows of short hairs from the collare to the scutellum; scutellum black or deep brown, levigate, with long hairs. Halteres brown, with short, sparse pubescence, the stalk sordid ochraceous. Abdomen black, with a moderately long, somewhat dense pubescence; lamellæ of the ovipositor black, elliptical. Legs pitch-brown. In the fore-legs the tarsi a little longer than the tibiæ; in the intermediate-legs the tibiæ somewhat longer than the tarsi; in the hind-legs the tibiæ about a quarter longer than the tarsi. Spurs honey-yellow, longer than the fourth tarsal joint. First joint of the tarsi rather more than twice the length of the second; second joint somewhat longer than the third; third and fifth joints of equal length, and one-third longer than the fourth. Wings pellucid, almost fuliginous, with the costal and two first longitudinal veins almost black; weak opaline reflections. First longitudinal vein reaching the costa immediately opposite the base of the fork; cross-vein very indistinct, situated immediately beyond the middle of the first longitudinal; petiole paler than the fork, shorter than the posterior branch; branches running almost parallel to one another, scarcely divergent at at their tips, the posterior branch very little arcuated. *fg* rather more than three times the length of *gh; kl* a little shorter than *lm.*

Hab.—North Waratah, near Newcastle (Skuse).   May.

B. Halteres yellow or whitish.

99. Sciara reciproca, Walker.

Sciara reciproca, Walker, Insecta Saundersiana, Vol. I. Diptera, 1856, p. 420.

" ♀.—*Nigra, obscura ; abdomen piceo-nigrum, thorace duplo longius ; pedes graciles, sat longi ; alæ cinereæ.*

" Black, dull. Abdomen piceous-black, about twice the length of the thorax. Legs slender, moderately long. Wings grey ; radial vein and cubital vein black ; the rest paler ; basal part of the subapical vein longer than its fork. Length of the body, $1\frac{1}{2}$ line ; of the wings, 3 lines.

" Van Diemen's Land."

Obs.—Without a re-description of this species it is impossible to say to what section or sub-section it may be referred, but Walker places it in the sub-division " b " of Meigen's division "A," which corresponds to its present position.

II. FIRST LONGITUDINAL VEIN JOINING THE COSTA BEFORE THE BASE OF THE FORK.

A. Halteres black or brown, the stalk wholly or partly yellow, yellowish or whitish.

1. Palpi black or brown.

B. Cross-vein situated at the middle of the first longitudinal vein.

b. *Tip of the second longitudinal vein and tip of the posterior branch of the fork equally near the apex of the wing.*

100. SCIARA FINITIMA, sp.n.

♀.—			
Length of antennæ	0·055 inch	...	1·39 millimètres.
Expanse of wings	0·140 × 0·050	...	3·55 × 1·27
Size of body	0·130 × 0·030	...	3·30 × 0·76

Antennæ pitch-brown, with a dense pale yellow pubescence ; not very slender, and not quite as long as the head and thorax together ; basal joints deep brown with a somewhat sparse, minute pubescence ; flagellar-joints sub-sessile, 2 to 3 times as long as broad. Head black. Eyes contiguous above. Palpi black. Thorax black, levigate, with three longitudinal rows of short yellowish hairs which extend from the collare almost to the scutellum, and a few long yellowish-brown hairs between the wings and the humeri; scutellum black, with a few long hairs on the posterior margin. Halteres pitch-brown, the base of the stalk ochraceous; club sparsely covered with short hairs. Abdomen ochraceous-brown, the dorsal segments laterally inclined to fuscous, and the four terminal segments with the lamellæ of the ovipositor almost umber brown; lamellæ almost elliptical. Legs pitch-brown. In the fore-legs the tarsi a little longer than the tibiæ; in the intermediate legs the tibiæ and tarsi of about equal length; in the hind-legs the tibiæ $\frac{1}{6}$ longer than the tarsi, and $\frac{1}{3}$ longer than the tibiæ of the intermediate-legs. Spurs yellowish, as long as the last joint of the tarsi. First tarsal joint in the two first pairs of legs rather more than twice the length of the second joint, in the hind-legs more than three times the length; second joint $\frac{1}{3}$ longer than the third, and about equal in length to the fourth and fifth joints together. Wings pellucid with a pale greyish-brown tint; brilliant margaritaceous reflections when viewed at a certain obliquity. First longitudinal vein joining the costa almost opposite but immediately before the base of the fork of the third longitudinal vein; petiole indistinct, and rather shorter than the anterior branch; anterior branch twice as arcuated at the base as the posterior; both branches scarcely divergent at the tips. *fg* about $1\frac{2}{3}$ times the length of *gh*; *kl* about the same length as *lm*.

Hab.—Glenbrook (Masters). End of November.

C. Cross-vein situated beyond the middle of the first longitudinal vein.

a. *Tip of the second longitudinal vein nearer the apex of the wing than the tip of the posterior branch of the fork.*

† *Thorax with two longitudinal rows of hairs.*

101. Sciara æmula, sp. n. (Pl. xi., fig. 3).

♀.—			
Length of antennæ	0·065 inch	...	1·66 millimètres.
Expanse of wings	0·180 × 0·065	...	4·56 × 1·66
Size of body	0·160 × 0·030	...	4·06 × 0·76

Antennæ black, with a short yellowish pubescence; rather longer than the head and thorax together; joints of the scapus with very little pubescence; flagellar joints sub-sessile, 2 to $2\frac{1}{2}$ times as long as broad, densely pubescent. Head black, levigate. Eyes almost contiguous above. Palpi deep brown, Thorax black, levigate, with two longitudinal and almost parallel rows of short brown hairs, extending almost to the scutellum, also some moderately long hairs along the lateral margins from the humeri; humeri slightly tipped with reddish-brown; scutellum with some moderately long hairs. Halteres entirely umber-brown, sprinkled with short hairs. Abdomen deep umber with the last two joints black, densely covered with a moderately long brown pubescence; as broad as the thorax; lamellæ of the ovipositor black, densely pubescent, elliptical. Coxæ black or very deep brown, with some rather long hairs on the front. Femora, tibiæ and tarsi deep umber-brown, densely pubescent. In the fore and intermediate legs the tarsi somewhat longer than the tibiæ; in the hind legs the tibiæ and tarsi of equal length. First joint of the tarsi 3 times the length of the second; second joint $\frac{1}{6}$ longer than the third and considerably shorter than the fourth and fifth together; third and fifth joints of about equal length, and longer than the fourth. Wings pellucid with a greyish-brown tint, the costal and two first longitudinal veins deep umber-brown; brilliant roseous and smaragdine reflections when viewed at a certain obliquity. First longitudinal vein reaching the costa almost opposite but immediately before the base of the fork of the third longitudinal; petiole paler than the fork and rather shorter than the posterior branch; posterior branch shorter than the anterior one, very little arcuated at the base; both branches slightly divergent at the tips. *fy* about twice the length of *gh*; *kl* about $\frac{3}{4}$ the length of *lm*.

Hab. – Elizabeth Bay and Middle Harbour (Skuse). September.

Obs.—In size, general appearance, and colour this species greatly resembles *S. Macleayi*, for which at first sight it might very easily be mistaken, but the length and narrowness of the fork of the third longitudinal vein (third sub-marginal cell) is at once a distinguishing character visible to the naked eye. I have only taken a single specimen in each of the above-named localites.

102. SCIARA LUCTIFICA, sp.n.

♂.—Length of antennæ	0·075 inch	1·89 millimètres.
Expanse of wings	0·140 × 0·050 ...	3·55 × 1·27
Size of body	0·120 × 0·025 ...	3·04 × 0·62

Antennæ pitch-brown, with a yellowish-brown pubescence; rather slender; more than half the length of the wings; basal joints pitch-brown, with a very sparse but longer pubescence than that on the flagellar joints; flagellar joints sub-sessile, 1½ to 2½ times as long as broad. Head black. Eyes contiguous above. Palpi pitch-brown. Thorax black or very deep brown, levigate with two indistinct longitudinal rows of very short yellowish-brown hairs; pleuræ deep reddish-brown; scutellum with a very minute sparse pubescence. Halteres pitch-brown, yellowish at the base, with a most minute sparse pubescence; club large, pyriform. Abdomen pitch-brown, with a somewhat dense long yellowish-brown pubescence; considerably broader at the base than the thorax, the last four segments becoming narrower; forceps pitch-brown, densely pubescent, broader than the terminal segment of the abdomen. Legs pitch-brown, densely covered with a fine yellowish-brown pubescence; fore femora rather shorter than the intermediate ones, hind femora somewhat longer than the latter; intermediate tibiæ ⅓ longer than the first, hind tibiæ slightly longer than the intermediate ones; spurs yellow, about as long as the fourth joint of the tarsi; tarsal joints of all the legs of about the same length, except that the first joint of the intermediate and hind tarsi is rather more than ⅙ longer than that of the fore tarsi; second joint of the tarsi ½ longer than the fourth; third joint just perceptibly longer than the fifth. Wings with a very pale somewhat reddish-brown tint, reflecting brilliant opaline colours when viewed at a certain obliquity. Petiole much paler than the fork and shorter than the anterior branch; tip of the anterior branch straight, tip of the posterior branch bent a little posteriorly. *fg* twice the length of *gh*; *kl* almost ¼ shorter than *lm*. Rudimentary fifth longitudinal vein close behind the fourth longitudinal vein disappearing at about ¾ of its length.

Hab.—Gawler, South Australia.

†† *Thorax with three longitudinal rows of hairs.*

103. SCIARA FROGGATTI, sp.n.

♂.—Length of antennæ...... 0·065 inch ... 1·66 millimètres.
Expanse of wings........ 0·110 × 0·045 ... 2·79 × 1·13
Size of body.............. 0·100 × 0·015 ... 2·54 × 0·38

Antennæ black or deep brown, with a yellowish pubescence; slender, about ⅔ the length of the body; joints of the scapus sparsely haired; flagellar joints sub-sessile, 2 to 3 times as long as broad, the terminal joint about ⅓ longer than the one immediately preceding it. Head black. Eyes not contiguous above. Palpi black. Thorax black, levigate, with three rows of short hairs extending from the collare almost to the scutellum; scutellum with a few moderately long hairs. Halteres black, somewhat brownish at the base of the stem, with a few very short hairs on the club. Abdomen black, slender, somewhat sparingly covered with a moderately long brownish pubescence; forceps large, wider than the abdomen, densely pubescent. Legs deep umber-brown, appearing almost black, particularly the tarsal joints. In the fore- and middle-legs the tarsi about ⅕ longer than the tibiæ; in the hind-legs the tibiæ and tarsi of about equal length. Spurs short, honey-yellow. First joint of the tarsi in the fore- and intermediate legs 3 times the length of the second, in the hind-legs rather more than twice the length; second joint ¼ longer than the third; third joint ⅓ longer than the fourth and a little longer than the fifth. Wings pellucid, almost fuliginous, with the costal and two first longitudinal veins nearly black; weak roscous and smaragdine reflections. First longitudinal reaching the costa some distance

before the base of the fork; cross-vein very distinct; petiole
paler than the fork, longer than the anterior branch; anterior
branch rather more arcuated than the posterior one; both
branches slightly divergent at the tips; branches of fourth
longitudinal vein darker than the last. *fg* twice the length of
gh; kl about equal in length to *lm.*

Hab.—Middle Harbour (Froggatt).   April.

104. Sciara frequens, sp.n.

♂.—Length of antennæ	0·060 inch	...	1·54 millimètres.
Expanse of wings	0·110 × 0·045	...	2·79 × 1·13
Size of body	0·100 × 0·015	...	2·54 × 0·38

Antennæ black, with a minute brownish-yellow pubescence;
rather slender, more than half the length of the body; basal
joints black, almost without any pubescence; flagellar joints with
very short pedicels, 2 to 4 times as long as broad, the terminal
joint considerably longer than the one immediately preceding it.
Head black.   Eyes almost contiguous above.   Palpi light
brown.   Thorax dull black or deep brown, with three longitudinal
and almost parallel rows of very minute golden-yellow single
hairs; scutellum with a short pubescence, the hairs longer than
those on the thorax.   Halteres light umber-brown, the base of the
stalk pale brownish-yellow, with sparse and minute pubescence;
club large, pyriform.   Abdomen deep umber-brown, with a very
sparse, short, golden-yellow pubescence; almost cylindrical, rather
narrower than the thorax; forceps very large, rather densely
pubescent.   Coxæ and femora very pale brownish-yellow, the
coxæ somewhat darker than the femora, the former with a few
short yellow hairs on the upper side, and the latter with a minute
pale pubescence also only on the upper side.   Trochanters light
brown, considerably darker than the coxæ and femora.   Tarsi
almost cinereous, but having a yellowish tinge.   In the fore-legs
the tarsi are about one-third longer than the tibiæ, in the intermediate-legs the tarsi are somewhat longer than the tibiæ, and in

the hind-legs the tibiæ are a very little longer than the tarsi. First tarsal joint nearly $2\frac{1}{2}$ times longer than the second, the second joint $\frac{1}{3}$ longer than the third, and as long as the fourth and fifth together. Wings cinereous, with brilliant pale green and rosy reflections. First longitudinal joining the costa immediately before the base of the fork of the third longitudinal vein ; petiole much paler than the fork, and considerably shorter than the anterior branch ; branches slightly divergent at the tips. fg $\frac{2}{3}$ longer than gh : kl somewhat shorter than lm.

Hab.—Sydney (Skuse). February.

b. *Tip of the second longitudinal vein and tip of the posterior branch of the fork equally near the apex of the wing.*

105. Sciara dolosa, sp.n.

♀.—Length of antennæ...... 0·045 inch ... 1·13 millimètres.

Expanse of wings........ 0·110 × 0·045 ... 2·79 × 1·13

Size of body............... 0·100 × 0·020 ... 2·54 × 0·50

Antennæ pitch-brown, with a yellow pubescence ; rather slender, about half the length of the body ; basal joints pitch-brown, with a very sparse pubescence ; flagellar joints sub-sessile, all but the terminal joint twice as long as broad, the latter one half longer than the preceding. Head black. Eyes almost contiguous above. Palpi dusky-brown. Thorax black, levigate, with three longitudinal rows of short brown hairs from the collare to the scutellum, also a patch of long hairs between the wings and the humeri ; scutellum with a somewhat sparse minute pubescence. Halteres dull pitch-brown, the base of the stem sordid ochraceous. Abdomen black on the dorsal segments, underneath umber-brown, sparsely pubescent ; in the middle about as broad as the thorax ; the lamellæ of the ovipositor black, oblong. Coxæ and femora testaceous, both darker on the upper side on account of their pubescence. Tibiæ and tarsi pitch-brown, the latter somewhat darker than the former, with a dense pubescence.

In the fore-legs the tarsi about ¼ longer than the tibiæ; in the intermediate-legs the tarsi and tibiæ of about equal length; and in the hind-legs the tibiæ somewhat longer than the tarsi. Spurs about the same length as the last joint of the tarsi, yellowish-brown. First tarsal joint rather more than twice the length of the second; the second ¼ longer than the third, and almost as long as the fourth and fifth together; fourth tarsal joint somewhat longer than the fifth. Wings almost hyaline, having a pale brownish tint, with brilliant reflections, in which blue predominates, when viewed at a certain obliquity. First longitudinal vein reaching the costa a very short distance before the base of the fork. Petiole rather paler than the fork and shorter than the anterior branch; posterior branch considerably less arcuated at the base than the anterior one, and almost in a line with the petiole; both branches very slightly divergent at the tips. *fg* rather more than twice the length of *gh*; *kl* a little shorter than *lm*.

Hab.—Elizabeth Bay, near Sydney (Skuse). December.

106. SCIARA FESTINA, sp.n.

♀.—Length of antennæ...... 0·040 inch   ...   1·01 millimètres.
  Expanse of wings....... 0·090 × 0·040 ... 2·27 × 1·01
  Size of body......... ..... 0·090 × 0·020 ... 2·27 × 0·50

Antennæ pitch-brown, with a pale yellow pubescence; slender, as long as the head and thorax together; basal joints dull castaneous, with a sparse but considerably longer pubescence than that of the flagellum; flagellar joints sub-sessile, twice as long as broad, with minute pedicels. Head black. Eyes contiguous above. Palpi yellowish-brown. Thorax black, levigate, with three double longitudinal rows of short hairs, nearly meeting just in front of the scutellum, also some long hairs between the wings and the humeri; humeri tipped with dull reddish-brown; scutellum black, with numerous short dispersed hairs. Halteres pitch-brown, the stem ochraceous, with a few minute hairs about the basal portion of the club. Abdomen somewhat dull castaneous, with a short pubescence; broader at the base and middle than the thorax; lamellæ of the ovipositor pitch-brown, elliptical. Legs pitch-brown. In the fore-legs the tarsi $\frac{1}{6}$ longer than the tibiæ; in the intermediate-legs the tarsi $\frac{1}{7}$ longer than the tibiæ; in the hind-legs the tibiæ somewhat longer than the tarsi. Spurs the same length as the last tarsal joint. First joint of the tarsi $2\frac{1}{2}$ times as long as the second, rather longer in the fore tarsi; second joint $\frac{1}{4}$ longer than the third and almost as long as the fourth and fifth joints together; two last joints of equal length. Wings pellucid, with a very pale greyish-brown tint, and having brilliant azure reflections when viewed at a certain obliquity. Veins pale yellowish brown. First longitudinal vein reaching the costa a short distance before the base of the fork; cross-vein somewhat ill-defined, slender; petiole most indistinct, and longer than the anterior branch of the fork; branches directed downwards, their tips slightly divergent; the posterior branch considerably shorter than the anterior. *fg* about $\frac{1}{2}$ longer than *gh; kl* almost the same as *lm.*

Hab.—Sydney (Masters and Skuse).

c. *Tip of the posterior branch of the fork nearer the apex of the wing than the tip of the second longitudinal vein.*

† *Thorax with two longitudinal rows of hairs.*

107. SCIARA PERNITIDA, sp. n.

♂.—Length of antennæ...... 0·100 inch ... 2·54 millimètres.
    Expanse of wings......... 0·100 × 0·035 ... 2 54 × 0·88
    Size of body .... 0·100 × 0·015 ... 2·54 × 0·38

♀.—Length of antennæ...... 0·060 inch ... 1·54 millimètres.
    Expanse of wings......... 0·140 × 0·045 ... 3·55 × 1·13
    Size of body........ ...... 0 120 × 0·025 ... 3·04 × 0·62

♂.—Antennæ slender, as long as the body, dark reddish-brown, with a minute pale pubescence; basal joints with very little pubescence; flagellar joints sub-sessile, 4 to 6 times as long as broad, the terminal joint considerably longer than the one immediately preceding it. Head black, sub-nitidous. Eyes non-contiguous above but very close. Palpi brown. Thorax black, very nitidous, with two longitudinal rows of minute hairs, slightly convergent, and not extending as far as the scutellum, also some long hairs between the humeri and the base of the wings ; scutellum deep brown, sub-nitidous, with two or three long hairs. Halteres pitch-brown, the base of the stem ochraceous. Abdomen black, with a short moderately dense pubescence, slender ; forceps considerably broader than abdomen, densely pubescent. Legs pitch-brown. In the fore-legs the tarsi rather more than $\frac{1}{4}$ longer than the tibiæ ; in the intermediate-legs the tarsi almost $\frac{2}{5}$ longer than the tibiæ ; and in the hind-legs the tarsi $\frac{1}{2}$ longer than the tibiæ. Spurs yellowish, as long as the last tarsal joint. First joint of the tarsi twice the length of the second ; second $\frac{1}{4}$ longer than the third joint, and rather longer than the fourth and fifth together ; fourth somewhat longer than the fifth. Wings pellucid, with a very pale somewhat reddish-brown tint and brilliant opaline reflections ; veins pale reddish-brown. First longitudinal vein reaching the costa considerably before the base of the fork : petiole very little paler than the fork, and almost as long as the anterior branch ; posterior branch very little arcuated at the base, and both slightly divergent at the tips. *fy* more than 4 times the length of *gh* ; *kl* rather longer than *lm*.

♀.—Antennæ half the length of the body, joints $2\frac{1}{2}$ to 3 times as long as broad. Abdomen robust ; lamellæ of the ovipositor brown, short, almost elliptical. Legs much darker than in the ♂. In the fore- and intermediate-legs the tarsi not quite $\frac{1}{4}$ longer than the tibiæ ; in the hind-legs the tarsi a little longer than the tibiæ.

Hab.—Elizabeth Bay (Masters and Skuse) ; Glenbrook, Blue Mountains (Masters). November.

Obs.—It is remarkable that we should have taken this species only in the two above-named and widely separate localities, where it occurs very abundantly.

108. Sciara familiaris, sp.n.

♂.—Length of antennæ...... 0·030 inch   ... 0·76 millimètre.
Expanse of wings....... 0·070 × 0·030 ... 1·77 × 0·76
Size of body.............. 0·070 × 0·012 ... 1·77 × 0·30

♀.—Length of antennæ...... 0·027 inch   ... 0.88 millimètre.
Expanse of wings....... 0·080 × 0·030 ... 2·02 × 0·76
Size of body........... ... 0·080 × 0·015 ... 2·02 × 0·38

♂.—Antennæ very slender, deep brown, with a dense pale pubescence ; joints of the scapus black, sparsely haired ; flagellar joints 1¼ to 1½ times as long as broad, five times as long as the pedicels. Head black. Eyes contiguous above. Palpi brown. Thorax black, levigate, with two longitudinal double rows of short sparse hairs running almost to the scutellum ; a few longer hairs between the origin of the wing and the humeri, also on the scutellum. Halteres umber brown, the base of the stem ochraceous-yellow ; short sparse pubescence on the club. Abdomen umber-brown, sordid ochraceous between the segments, with a moderately long, rather dense, pubescence; forceps and terminal segment almost black, the former wider than the two segments immediately preceding them, very densely haired. Legs greyish-brown. In the fore- and intermediate-legs the tarsi a little longer than the tibiæ; in the hind-legs the tibiæ slightly longer than the tarsi. Spurs as long as the fourth tarsal joint. Fourth joint of the tarsi 2½ times the length of the second ; second joint ¼ longer than the third and almost equal in length to the fourth and fifth combined ; third joint somewhat longer and considerably thicker than the fifth, and about ⅓ longer than the fourth. Wings almost hyaline, with a very pale greyish-brown tint; veins light umber-brown; brilliant violaceous and purpureous reflection when viewed at a certain obliquity. First longitudinal vein reaching the costa a short distance before the root of the fork, and opposite to the tip of the posterior branch of the fourth longitudinal ; cross-vein very distinct, a little beyond the middle of the first longitudinal vein ; petiole more indistinct than the fork, rather shorter

than the posterior branch; branches running parallel for the
greater part of their length, scarcely divergent at the tips, the
posterior branch very little arcuated at the base, and almost
straight; branches of the fourth longitudinal vein considerably
more distinct than the third longitudinal. *fg* three times the
length of *gh*; *kl* almost imperceptibly shorter than *lm*.

♀.—Antennæ about as long as the head and thorax together;
flagellar joints sub-sessile, $1\frac{1}{4}$ to $2\frac{1}{4}$ times as long as broad, the
pedicels very minute; terminal joint longer and thicker than the
preceding; joints of the scapus brown. Abdomen wider than the
thorax, almost cylindrical, uniform umber-brown, rather sparsely
pubescent; lamellæ of the ovipositor small, oval, black or deep
brown, with a very short dense pubescence. Branches of the fork
of the third longitudinal vein not running so parallel to one-another
as in the ♂ but gradually divergent.

Hab.—Elizabeth Bay (Skuse). January.

103. Sciara cavatica, sp.n.

♂.—Length of antennæ...... 0·030 inch   ...   0·76 millimètre.
Expanse of wings........ 0·075 × 0·025 ...   1·89 × 0·62
Size of body............. 0·065 × 0·015 ...   1·66 × 0·38

Antennæ pitch-brown, with a bright yellowish-brown pubescence; slender, about half the length of the body; basal
joints rather paler than those of the flagellum, and not so thickly
haired; flagellar joints sub-sessile, the pedicels more distinctly
visible towards the tip where they are about $\frac{1}{4}$ to $\frac{1}{3}$ the
length of the joints; all but the terminal joint about one-half
longer than broad, the latter being about one-half longer than
the preceding; pilose. Head black. Eyes contiguous above.
Thorax black, or deep brown, levigate, with two longitudinal
double rows of short yellowish-brown hairs; humeri slightly
tipped with ochraceous; scutellum pitch-brown, with some
yellowish reflections on the posterior border, sparsely covered
with a moderately long yellowish pubescence. Halteres pitch
brown, ochraceous at the base of the stem, with a minute

pubescence. Abdomen umber-brown, moderately clothed with
yellowish hairs ; at the base rather narrower than the thorax,
gradually dilated towards the middle ; forceps large, densely
haired. Legs ochraceous-brown ; the coxæ, tibiæ and tarsi densely
pubescent ; femora darker on the upper side on account of their
pubescence. In the fore-legs the tarsi about $\frac{1}{5}$ longer than
the tibiæ ; in the intermediate-legs both of equal length ; in
the hind-legs the tibiæ about $\frac{1}{3}$ longer than the tarsi. Spurs
ochraceous-brown, as long as the fourth tarsal joint. First tarsal
joint three times the length of the second ; second $\frac{1}{4}$ longer than
the third ; third and fifth the same length and $\frac{1}{5}$ longer than
the fourth joint. Wings pellucid, with a pale yellowish-brown
tint, and brilliant azure and violaceous reflections when viewed at
a certain obliquity. Costal and first and second longitudinal veins
very distinct, yellowish-brown ; the rest of the veins indistinct.
First longitudinal vein reaching the costa a very short distance
before the base of the fork ; cross-vein very thick ; petiole rather
shorter than the posterior branch of the fork ; branches not
divergent at their tips ; tip of the posterior branch very little
nearer the apex of the wing than the tip of the second
longitudinal vein. *fg* four times the length of *gh* ; *kl* almost the
same as *lm*.

Hab. — Glenbrook, Blue Mountains (Masters). End of
November.

110. SCIARA FESTIVA, sp.n.

♂.—Length of antennæ... .. 0·030 inch ... 0·76 millimètre.
Expanse of wings........ 0·070 × 0·030 ... 1·77 × 0·76
Size of body.............. 0·060 × 0·012 ... 1·54 × 0·30

♀.—Length of antennæ...... 0·025 inch ... 0·62 millimètre.
Expanse of wings........ 0·070 × 0·030 ... 1·77 × 0·76
Size of body.............. 0·065 × 0·015 ... 1·66 × 0·38

♂.—Antennæ slender, half the length of the body, umber-
brown with a short dense yellowish pubescence ; joints of the
scapus sparsely pubescent ; flagellar joints sub-sessile, towards the

tip about one-half longer than broad, the terminal joint more slender, about $3\frac{1}{2}$ times as long as broad. Head black, levigate. Eyes contiguous above. Palpi light brown. Thorax black, levigate, with two indistinct longitudinal double rows of very short hairs reaching nearly to the scutellum; scutellum deep reddish-brown, with a short stiff pubescence. Halteres brown, the stalk brownish-yellow, club with a few scattered very short hairs. Abdomen broader than the thorax, black, rather densely clothed with a moderately long pubescence; forceps black, densely pubescent, a little wider than the abdomen. Legs ochraceous-brown, the tibiæ and tarsi darker than the rest. In the fore-legs the tarsi a very little longer than the tibiæ; in the intermediate legs the tibiæ almost imperceptibly longer than the tarsi; and in the hind-legs the tibiæ nearly $\frac{1}{6}$ longer than the tarsi. Spurs about as long as the fourth tarsal joint. First joint of the tarsi in the first and second pairs of legs about $2\frac{1}{2}$ times the length of the second, in the hind-legs 3 times the length; second joint somewhat longer than the third; third joint somewhat longer than the fifth. Wings pellucid, with a very pale yellowish-brown tint; brilliant blue and purpureous reflections. First longitudinal vein reaching the costa a short distance before the root of the fork; cross-vein very indistinct; third longitudinal vein almost uniformly indistinct, the petiole shorter than the lower branch of the fork; branches running almost parallel for the quarter part of their length, scarcely divergent at the tips; the posterior branch very little arcuated at the base; fork of the fourth longitudinal vein more distinct than the last. fy almost $3\frac{1}{2}$ times the length of gh; kl about equal to lm.

♀.—Antennæ slender, as long as the head and thorax together, the pubescence short, not quite so dense as in the ♂; flagellar joints very little longer than broad, the terminal joint about 3 times as long as wide. Lamellæ of the ovipositor small, black, densely pubescent, ovate.

- *Hab.*—Elizabeth Bay (Skuse). May.

††. *Thorax with three longitudinal rows of hairs.*

111. Sciara mæsta, sp.n.

♂.—Length of antennæ	......	0·100 inch	...	2·54 millimètres.
Expanse of wings	....	0·120 × 0·45	...	3·04 × 1·13
Size of body	....	0·100 × 0·015	...	2·54 × 0·38

♀.—Length of antennæ	......	0·060 inch	...	1·54 millimètres.
Expanse of wings	....	0·160 × 0·055	...	4·06 × 1·39
Size of body	....	0·140 × 0·030	...	3·55 × 0·76

♂.—Antennæ somewhat thick, deep brown, with a long dense yellowish pubescence ; basal joints black, with a sparse pubescence ; flagellar joints sub-sessile $1\frac{1}{2}$ to 3 times as long as broad, with short pedicels. Head black, levigate. Eyes almost contiguous above. Palpi deep brown or black. Thorax black, levigate, with three longitudinal rows of very short hairs, the lateral ones slightly convergent, and running almost as far as the scutellum, the intermediate one only visible when viewed at a certain obliquity, reaching to the middle of the thorax ; also some short hairs between the base of the wings and the humeri ; humeri tipped with deep reddish-brown ; scutellum deep brown almost black, with a very sparse short pubescence. Halteres greyish-brown, with a few short hairs, the stem yellow. Abdomen slender, black, with moderately long golden-yellow hairs, the pubescence considerably more dense on the last three or four segments ; forceps densely haired, very little if any wider than the abdomen. Fore coxæ honey-yellow, dusky at the base ; middle and hind coxæ deep brown or black. Femora honey-yellow. Tibiæ and tarsi darker, partly on account of their pubescence, almost pitch-brown. In the fore- and intermediate-legs the tarsi $\frac{1}{4}$ longer than the tibiæ ; in the hind-legs the tibiæ are slightly longer than the tarsi. Spurs yellow, about equal in length to the fourth tarsal joint. First joint of the tarsi more than twice the length of the second ; second joint $\frac{1}{3}$ longer than the third, and about as long as the fourth and fifth together ; fifth joint some-what longer than the fourth. Wings pellucid, with a pale greyish-brown tint ; opaline reflections. Costal and two first longitudinal

45

veins brown ; very distinct. Cross-vein thick. First longitudinal vein reaching the costa some distance before the base of the fork ; petiole somewhat paler than the fork and longer than the anterior branch ; posterior branch shorter than the anterior and very little arcuated at its base ; tips scarcely divergent. *fg* three times the length of *gh ; kl* a little longer than *lm*.

♀.—Antennæ rather slender, pitch-brown, with a moderately long dense yellowish pubescence ; basal joints pitch-brown, somewhat paler than those of the flagellum, with a sparse pubescence ; flagellar joints 2 to 4 times as long as broad, sub-sessile. Halteres brown, with a few hairs, the stem brownish-yellow. Abdomen deep brown or black ; lamellæ of the ovipositor deep brown or black, elliptical. First joint of the tarsi 3 times the length of the second ; second joint about ¼ longer than the third, and not quite the length of the combined fourth and fifth joints ; fifth joint about ⅓ longer than the fourth.

Hab.— Middle Harbour, near Sydney (Skuse); Berowra (Masters). August.

112. SCIARA AQUILA, sp.n.

♀.—Length of antennæ...... 0·040 inches   ...   1·01 millimètres.
Expanse of wings........ 0·140 × 0·050 ... 3·55 × 1·27
Size of body...... .... 0·120 × 0·020 ... 3·04 × 0·50

Antennæ black, with a short yellowish pubescence ; slender, as long as the head and thorax combined ; joints of the scapus black, sparsely pubescent ; flagellar joints sub-sessile, 2 to 2½ times as long as broad. Head black. Eyes contiguous above. Palpi deep brown. Thorax black, sub-nitidous, with three longitudinal rows of golden-yellow pubescence, the lateral rows starting from the humeri, extending almost to the scutellum and nearly meeting, the hairs becoming less dense towards the scutellum ; the intermediate row very short and indistinct, sparsely haired ; a few lateral hairs in front of the origin of the wings, not extending to the humeri ; humeri very slightly tipped with reddish-brown ; scutellum with a few moderately long, golden-yellow hairs. Halteres almost fuscous with the stem pitch-brown, with no visible pubescence. Abdomen umber-brown, appearing almost black in a certain light, with a very short pubescence; lamellæ of the ovipositor umber-brown, elliptical. Legs pitch-brown. In the fore-legs the tarsi almost imperceptibly longer than the tibiæ; in the intermediate legs the tibiæ a little longer than the tarsi; in the hind-legs the tibiæ $\frac{1}{5}$ longer than the tarsi. Spurs yellowish, equal in length to the fourth tarsal joint. First joint of the tarsi 3 times the length of the second; second joint $\frac{1}{4}$ longer than the third and shorter than the fourth and fifth combined; third and fifth joints about the same length and $\frac{1}{4}$ longer than the fourth. Wings pellucid, with a pale greyish-brown tint with brilliant roseous and smaragdine reflections. Costal and first two longitudinal veins somewhat reddish-brown. Cross-vein very distinct, narrower than the second longitudinal. First longitudinal vein joining the costa almost opposite the base of the fork of the third longitudinal vein; petiole paler than the fork, and shorter than than its posterior branch; posterior branch less arcuated at the base than the anterior one; branches running almost parallel, slightly divergent at their tips; tip of the posterior branch very little nearer the apex of the wing than the tip of the second longitudinal vein. *fg* twice the length of *gh;* *kl* about $\frac{2}{3}$ the length of *lm.*

Hab.—Glenbrook (Masters). November.

Obs.—I have only seen two specimens.

113. Sciara audax, sp.n.

♀.—Length of antennæ...... 0·050 inch ... 1·27 millimètres.
  Expanse of wings....... 0·110 × 0·040 ... 2·79 × 1·01
  Size of body .... 0·110 × 0·020 ... 2·79 × 0·50˙

Antennæ slender, nearly half the length of the body, black, with a minute pale pubescence; joints of the scapus black, sparsely pubescent; flagellar joints sub-sessile, about $2\frac{1}{2}$ times as long as broad, the terminal joint longer. Head black. Eyes contiguous above. Palpi black or very deep brown. Thorax black, levigate, with three longitudinal rows of golden-yellow hairs, the intermediate one a single row of short hairs, rather indistinct, reaching almost as far as the other two, the lateral rows double, with somewhat longer hairs extending almost to the scutellum ; also some long golden-yellow hairs just anterior to the origin of the wings; scutellum sparsely covered with long golden-yellow hairs. Halteres deep pitch-brown, with a very short sparse pubescence, the stem paler. Abdomen wider than the thorax, deep pitch-brown, somewhat paler between the segments, covered with a very short yellowish pubescence ; lamellæ of the ovipositor small, oval. Coxæ deep pitch-brown, ferruginous at the apex, and with some rather long yellowish hairs on the upper side. Femora deep pitch-brown with a tolerably dense, short pubescence. Tibiæ and tarsi almost fuliginous on account of their dense pubescence. In the fore-legs the tarsi about $\frac{1}{6}$ longer than the tibiæ ; in the intermediate legs the tarsi somewhat longer than the tibiæ ; and in the hind-legs the tibiæ and tarsi are of about equal length. Spurs about the length of the fourth tarsal joint. First joint of the tarsi about $2\frac{1}{2}$ times the length of the second ; second joint nearly $\frac{1}{4}$ longer than the third ; third joint $\frac{1}{3}$ longer than the fourth, and somewhat longer than the fifth joint. Wings pellucid with an almost fuliginous tint ; costal and two first longitudinal veins obscure brown ; brilliant margaritaceous reflections. First longitudinal vein reaching the costa a short distance before the base of the fork and a little beyond the tip of the posterior branch of the fourth longitudinal vein ; cross-vein distinct, a short distance beyond the middle of the first longitudinal ; petiole very pale and indistinct, shorter than the posterior branch of the fork ; branches running almost parallel for the greater part of their length, slightly divergent at the tips ; the posterior branch very little arcuated at the base. fg nearly $3\frac{1}{2}$ times the length of gh; kl shorter than lm.

Hab.—Elizabeth Bay (Masters). May.

114. Sciara vecors, sp.n.

♀.—Length of antennæ...... 0·050 inch ... 1·27 millimètres.
Expanse of wings........ 0·100 × 0·030 ... 2·54 × 0·76
Size of body .. .. 0·080 × 0·015 ... 2·02 × 0·38

Antennæ rather slender, black, with a minute, very dense, pale pubescence ; joints of the scapus sparsely haired ; flagellar joints sub-sessile, 2 to 2½ times as long as broad. Palpi black or deep brown. Eyes contiguous above. Head black, sub-nitidous. Thorax black, sub-nitidous, with three longitudinal double rows of short pale hairs, the intermediate one indistinct, reaching to the middle of the thorax, the lateral ones running almost to the scutellum, also some long hairs between the scutellum and the humeri ; scutellum with some long hairs. Halteres deep brown, the stem paler brown, with apparently no pubescence. Abdomen black, rather sparsely covered with a short pale pubescence ; lamellæ of the ovipositor black, oval. Legs obscure pitch-brown, the coxæ almost black. In the fore-legs the tarsi are a little longer than the tibiæ ; in the intermediate-legs the tibiæ and tarsi are of almost equal length ; and in the hind-legs the tibiæ are a very little longer than the tarsi. Spurs as long as the fifth tarsal joint. First joint of the tarsi about 2½ times the length of the second ; second joint almost ¼ longer than the third and nearly equal to the fourth and fifth together ; third joint ⅓ longer than the fourth and somewhat longer than the fifth. Wings almost hyaline, with a very pale greyish-brown tint, the veins light umber-brown ; brilliant rosy and blue margaritaceous reflections when viewed at a certain obliquity. First longitudinal vein reaching the costa a short distance before the base of the fork and a little beyond the tip of the posterior branch of the fourth longitudinal vein ; petiole considerably paler than the fork, and about equal in length to the anterior branch ; branches running almost parallel, very slightly divergent at their tips ; posterior branch very little arcuated at the base. *fg* rather more than 2⅓ times the length of *gh* ; *kl* about ¾ the length of *lm*.

Hab.—Tenterfield, New England (Skuse).   February.

Obs.—The only specimen I have seen of this I took from a cobweb.

115. SCIARA ERRATICA, sp.n.

♀.—Length of antennæ...... 0·030 inch   ...   0·76 millimètre.
    Expanse of wings........ 0·085 × 0·030 ...   2·14 × 0·76
    Size of body.............. 0·090 × 0·015 ...   2·27 × 0·38

Antennæ slender, one-third the length of the body, black or deep brown, with a short dense yellowish pubescence ; joints of the scapus brown, with a very short sparse pubescence; flagellar joints sub-sessile, 2 to 2½ times as long as broad. Head black, with a short, rather sparse, yellowish pubescence. Eyes contiguous above. Palpi brown. Thorax black, levigate, with three longitudinal double rows of yellowish hairs, coalescent at the scutellum; a few lateral hairs before the humeri ; scutellum black with some short yellowish hairs. Halteres pale brown, ochraceous at the base, with a few very short hairs. Abdomen deep umber-brown, pale between the segments, sparsely clothed with a moderately long yellowish pubescence ; lamellæ of the ovipositor very small, umber-brown, elliptical, not very densely pubescent. Coxæ honey-yellow, with a few longish yellow hairs. Femora, tibiæ and tarsi ochraceous-brown, the two latter, particularly the terminal joints of the tarsi, darker on account of their dense pubescence. In the fore-legs the tarsi about ⅓ longer than the tibiæ ; in the intermediate-legs the tibiæ almost imperceptibly longer than the tarsi ; in the hind-legs the tibiæ only a little longer than the tarsi. Spurs very short. First joint of the tarsi 2½ times the length of the second ; second joint ¼ longer than the third ; third joint about ⅓ longer than the fourth and a little longer than the fifth. Wings almost hyaline, with a yellowish tint, margaritaceous reflections when viewed at a certain obliquity. Veins yellowish-brown. First longitudinal vein reaching the costa considerably before the base of the fork, and about opposite to the tip of the posterior branch of the fourth longitudinal vein ; cross-vein somewhat indistinct ; petiole almost invisible, somewhat shorter than the anterior branch of the fork ; branches of the fork running almost parallel, a little divergent at their tips, the posterior branch scarcely arcuated at its base. *fg* about ¼ longer than *gh ; kl* rather shorter than *lm.*

Hab.—Hexham (Skuse). April.

116. SCIARA APPROXIMATA, sp.n.

♀.—Length of antennæ...... 0·040 inch ... 1·01 millimètres.
  Expanse of wings....... 0·080 × 0·030 ... 2·02 × 0·76
  Size of body.............. 0·075 × 0·015 ... 1·89 × 0·38

Antennæ slender, deep brown, longer than the head and thorax together, with a short, dense, pale yellowish pubescence ; joints of the scapus deep brown, sparsely, haired ; flagellar joints sub-sessile, twice as long as broad. Head black, nitidous, sparsely pubescent. Eyes non-contiguous but very close. Palpi ochraceous-brown. Thorax black, nitidous, with three indistinct longitudinal double rows of yellowish hairs, the lateral ones reaching almost to the scutellum ; also a few long hairs between the origin of the wings and the humeri, and on the scutellum. Halteres deep brown, almost black, ochraceous at the base of the stalk, club sprinkled with some very short hairs at the base. Abdomen very deep umber-brown, of a lighter shade between the segments, with a tolerably dense pubescence ; lamellæ of the ovipositor very small, brown, elliptical, with brownish-yellow pubescence. Coxæ and femora ferruginous-ochraceous. Tibiæ and tarsi dusky-brown. In the fore-legs the tarsi a little longer than the tibiæ ; in the intermediate-legs the tibiæ and tarsi of equal length ; in the hind-legs the tibiæ a little longer than the tarsi. Spurs very short. First joint of the tarsi 3 times the length of the second ; second joint a little longer than the third ; third and fifth joints of about equal length, and ⅓ longer than the fourth. Wings pellucid, with a pale yellowish-brown tint ; brilliant margaritaceous reflections. Veins yellowish-brown. First longitudinal vein reaching the costa a short distance before the base of the fork ; cross-vein thick ; petiole very pale, almost as long as the anterior branch of the fork ; branches running almost parallel, a little divergent at the tips ; posterior branch very little arcuated at the base. *fg* about 5 times the length of *gh; kl* shorter than *lm*.

Hab.—Sydney (Skuse). January.

117. Sciara evanescens, sp.n.

♀.—Length of antennæ...... 0·030 inch ... 0·76 millimètre.
Expanse of wings........ 0·080 × 0·030 ... 2·02 × 0·76
Size of body .... 0·075 × 0·015 ... 1·87 × 0·38

Antennæ slender, not as long as the head and thorax together, umber-brown, with a short, dense, yellowish pubescence ; joints of the scapus lighter brown, sparsely pubescent ; flagellar joints sub-sessile, 1¼ times as long as broad, the terminal joint very slender and almost twice the length of the joint immediately preceding it. Head black, levigate. Eyes contiguous above. Palpi light brown. Thorax deep brown, almost black, levigate, with three longitudinal double rows of brownish-yellow hairs, the intermediate row very indistinct, stopping before the middle, the lateral ones sparse, not extending quite to the scutellum and not coalescent ; some long hairs between the origin of the wings and the humeri, also on the scutellum. Halteres deep brown, the base of the stalk brownish-yellow ; a few very short hairs about the base of the club. Abdomen deep umber-brown, almost black on the last few segments, rather densely clothed with a moderately long pale pubescence ; lamellæ of the ovipositor small, deep brown, oval. Legs pitch-brown, the tibiæ, and particularly the tarsi, darker on account of their dense pubescence. The tibiæ and tarsi of the fore-legs short, of equal length ; in the intermediate-legs the tibiæ and tarsi a little longer than the last, the tibiæ being about ½ longer than the tarsi ; in the hind-legs the tibiæ ⅓ longer than the tarsi. Spurs as long as the fourth tarsal joint. First joint of the tarsi in the fore- and intermediate-legs

2½ times, and in the hind-legs 3 times the length of the second ; second joint a little longer than the third ; third and fifth joints of about equal length and ¼ longer than the fourth. Wings pellucid, with a pale yellowish-brown tint; reflecting brilliant æneous and chalybeous tints when viewed at a certain obliquity. First longitudinal vein reaching the costa a little before the base of the fork ; petiole very indistinct, somewhat shorter than the anterior branch of the fork ; branches pale, running almost parallel, slightly divergent at the tips, the posterior branch very little arcuated at the base. *fg* about 4 times the length of *gh ; kl* a little shorter than *lm*.

Hab.—Sydney (Skuse). December.

118. Sciara scitula, sp.n.

♀.—Length of antennæ...... 0·035 inch ... 0·88 millimètre.
Expanse of wings........ 0·080 × 0·030 ... 2·02 × 0·76
Size of body .... 0·070 × 0·015 ... 1·77 × 0·38

Antennæ black or deep brown, with a minute pale pubescence; slender, as long as the head and thorax combined ; joints of the scapus with a very sparse and minute pubescence ; flagellar joints sub-sessile, 2 to 2½ times as long as broad, the terminal joint nearly 4 times as long. Head black, sub-nitidous. Eyes almost contiguous above. Palpi reddish-brown. Thorax black, sub-nitidous, with three longitudinal double rows of yellowish-brown hairs, the intermediate one somewhat indistinct, reaching only to the middle of the thorax, the lateral ones running almost as far as the posterior margin ; also a row of longer hairs on the lateral margins between the origin of the wings and the humeri ; scutellum with a few moderately long hairs. Halteres pitch-brown, a few short hairs on the club. Abdomen deep brown appearing almost black, with a short sparse yellowish-brown pubescence, dense at the extremity ; lamellæ of the ovipositor deep brown, small, oval. Coxæ honey-yellow. Femora, tibiæ and tarsi pitch-brown. In the fore- and intermediate-legs the tarsi somewhat longer than the tibiæ ; in the hind-legs the tibiæ a little longer than the tarsi. Spurs shorter than the fourth tarsal joint. First joint of the tarsi rather more than twice the length of the second ; second joint longer than the third ; third and fifth joints of about equal length and about ⅓ longer than the fourth. Wings pellucid, almost hyaline, with a very pale yellowish-brown tint ; brilliant blue and purple reflections when viewed in a certain light. Veins pale yellowish-brown. First longitudinal vein joining the costa a short distance before the base of the fork ; cross-vein distinct, considerably beyond the middle of the first longitudinal ; petiole very pale and indistinct, about the same length as the anterior branch of the fork ; branches gradually separating as they proceed to the margin, and slightly divergent at their tips ; posterior branch very little arcuated, and very little nearer the apex of the wing than the tip of the second longitudinal vein. *fg* about 2½ times longer than *gh ; kl* rather shorter than *lm*

Hab.—Sydney (Skuse).

119.—SCIARA BREVIFURCA, sp.n.

♀.—Length of antennæ...... 0·030 inch   ...   0·76 millimètre.
Expanse of wings........ 0·055 × 0·025 ...   1·39 × 0·62
Size of body.............. 0·050 × 0·0:5 ...   1·27 × 0·38

Antennæ slender, rather more than half the length of the body, deep brown, almost black, with a short dense yellowish pubescence ; joints of the scapus umber-brown, sparsely pubescent ; flagellar joints a little longer than broad, the terminal joint nearly twice the length of the one immediately preceding it. Head black, levigate. Eyes almost contiguous. Palpi light umber-brown. Thorax deep brown, levigate, with three longitudinal double rows of short yellowish-brown hairs, the intermediate one very indistinct reaching only to about the middle, the lateral ones extending to the scutellum, but not coalescent ; some moderately long hairs between the origin of the wings and the scutellum, also a few on the scutellum ; scutellum umber-brown. Halteres umber-brown, the base of the stalk yellowish, sprinkled with a few short hairs on the club. Abdomen faded olivaceous-brown on the dorsal segments, paler between the segments and underneath, sparsely clothed with a short pubescence; ovipositor dusky-brown, the lamellæ small, elliptical. Legs faded olivaceous-brown. In the fore-legs the tibiæ and tarsi are very short, the tarsi a little longer than the tibiæ; in both the intermediate- and hind-legs the tarsi somewhat longer than the tibiæ. Spurs very minute. First joint of the tarsi twice the length of the second; second joint somewhat longer than the third; third and fifth joints of equal length, each about $\frac{1}{4}$ longer than the fourth. Wings hyaline, the veins tinted with pale brownish; cupreous reflections when viewed at a certain obliquity. First longitudinal vein reaching the costa some distance before the base of the fork, and opposite to the tip of the posterior branch of the fourth longitudinal fork; cross-vein indistinct; third longitudinal vein, with its fork, indistinct; the petiole longer than the fork; fork very angular at the base, gradually widening to the margin of the wing, tips somewhat divergent; in some specimens the fork is most indistinct at the base; branches of the fourth longitudinal vein more distinct than the third longitudinal. *fy* rather more than twice the length of *gh*; *kl* somewhat shorter than *lm*.

Hab.—Elizabeth Bay (Skuse). January to May.

120. SCIARA DIVERSA, sp.n.

♀.—Length of antennæ	0·025 inch	...	0·62 millimètre.
Expanse of wings	0·060 × 0·025	...	1·54 × 0·62
Size of body	0·060 × 0·015	...	1·54 × 0·38

Antennæ very slender, nearly half the length of the body, deep umber-brown, with a short dense yellowish pubescence; joints of the scapus with a very short sparse pubescence; flagellar joints sub-sessile, near the tip about $\frac{1}{4}$ longer than broad, the terminal joint rather more than twice as long as broad. Head black. Eyes almost contiguous above. Palpi brown. Thorax black, sub-nitidous, with three longitudinal double rows of short pale brownish hairs from the collare to the scutellum, also some long hairs between the origin of the wings and the scutellum ; scutellum black, sub-nitidous, sparsely covered with a short pubescence, with two or three very long setæ. Halteres deep brown, the base of the stem almost ochraceous-brown, with no perceptible pubescence. Abdomen black or very deep brown, with a sparse covering of short, pale brownish, pubescence ; lamellæ of the ovipositor rather small, ovate. Legs light pitch-brown. In the fore-legs the tibiæ and tarsi very short, the tarsi a little longer than the tibiæ ; in the intermediate-legs the tibiæ and tarsi of about equal length ; in the hind-legs the tibiæ about $\frac{1}{7}$ longer than the tarsi. Spurs about the same length as the fourth tarsal joint. First joint of the tarsi about $2\frac{3}{4}$ times as long as the second ; second joint $\frac{1}{3}$ longer than the third ; third joint $\frac{1}{3}$ longer than the fourth, and somewhat longer than the fifth. Wings hyaline, with brown veins and golden and roseous reflections. First longitudinal vein reaching the costa some distance before the base of the fork and abou opposite the tip of the posterior branch of the fourth longitudinal fork ; cross-vein distinct ; petiole paler than the fork, and shorter than either branch ; fork not pointed at the base, the branches running almost parallel for the greater part of their length, but divergent towards their tips, posterior branch slightly arcuated at the base ; branches of the fourth longitudinal vein rather more distinct than the last. *fg* about $\frac{1}{5}$ longer than *gh;* *kl* about $\frac{3}{8}$ the length of *lm*.

Hab.—Elizabeth Bay (Masters). May.

121. SCIARA MINUTELA, sp.n.

♂.—Length of antennæ...... 0·030 inch   ...   0·76 millimètre.
    Expanse of wings........ 0·050 × 0·020 ...   1·27 × 0·50
    Size of abdomen .... 0·050 × 0·010 ...   1·27 × 0·25

♀.—Length of antennæ...... 0·026 inch   ...   0·62 millimètre.
    Expanse of wings........ 0·055 × 0·020 ...   1·39 × 0·50
    Size of abdomen .... 0·050 × 0·010 ...   1·27 × 0·25

♂.—Antennæ slender, more than half the length of the body, deep umber-brown, with a short dense yellowish pubescence ; joints of the scapus deep brown, rather sparsely pubescent ; flagellar joints sub-sessile, twice as long as broad, the terminal joint one-half longer than the one immediately preceding it. Head black. Eyes contiguous above. Palpi yellow. Thorax black, levigate, with three longitudinal double rows of short yellowish hairs, the intermediate one stopping beyond the middle of the thorax, the lateral ones reaching the scutellum ; a few hairs anterior to the origin of the wings and on the scutellum. Halteres deep brown, the base of the stalk somewhat paler, with a few short hairs. Abdomen black, or very deep brown, nearly as wide as the thorax, covered with short yellowish hairs ; forceps not as wide as the abdomen, densely pubescent. Legs brownish-ochraceous. In the fore-legs the tarsi very slightly longer than the tibiæ ; in the intermediate-legs the tarsi $\frac{1}{10}$ longer than the tibiæ ; in the hind-legs the tibiæ almost imperceptibly longer than the tarsi. Spurs as long as the fourth tarsal joint. First joint of the tarsi $2\frac{1}{2}$ times the length of the second ; second joint a little longer than the third ; third joint $\frac{1}{4}$ longer than the fourth and slightly longer than the fifth joint. Wings almost hyaline, with a more or less brilliant aurichalceous reflection when viewed at a certain obliquity. Veins yellowish-brown, the third and fourth longitudinals pale. First longitudinal vein joining the costa considerably before the base of the fork, and somewhat before the tip of the posterior branch of the fourth longitudinal ; cross-vein pale ; petiole scarcely visible, longer than the anterior branch of the fork ; branches little arcuated, particularly the anterior one making the fork rather angular at the base ; tips distinctly divergent. _fy_ exactly the same length as _gh_ ; _kl_ a little shorter than _lm_.

♀.—Antennæ very slender, almost half the length of the body, with a very short dense pubescence ; flagellar joints one-half longer than broad ; terminal joints twice as long as broad. Head, thorax and abdomen deep brown ; lamellæ of the ovipositor small, oval. Petiole scarcely paler than the fork, shorter than the anterior branch, about the same length as the posterior branch. *fg* nearly twice the length of *gh*.

Hab.—Glenbrook, Blue Mountains (Masters). November.

122. SCIARA ATRATULA, sp.n.

♀.—Length of antennæ...... 0·020 inch   ...   0·50 millimètre.
Expanse of wings....... 0·050 × 0·020 ...   1·27 × 0·50
Size of body.............. 0 045 × 0·010 ...   1.13 × 0·25

Antennæ slender, as long as the head and thorax together, black, with a dense brown pubescence; joints of the scapus brown, with apparently no pubescence; flagellar joints sub-sessile, rather longer than broad towards the tip, the terminal joint about twice as long as broad. Head black. Eyes almost contiguous above. Palpi black or deep brown. Thorax black, levigate, with three very indistinct longitudinal double rows of very short brownish hairs, the intermediate one particularly indistinct, apparently only extending to the middle of the thorax, the lateral ones wider, running almost to the scutellum; a few long brownish hairs anterior to the origin of the wings; apparently no pubescence on the scutellum. Halteres black, the base of the stem sordid ochraceous; club very little thickened; apparently no pubescence. Abdomen rather wider than the thorax, deep olive-brown, darker between the segments, clothed with a very short dense pubescence; lamellæ of the ovipositor very small, elongate. Coxæ and femora fuliginous-ochraceous. Tibiæ and tarsi almost fuliginous. In the fore-legs the tarsi nearly $\frac{1}{6}$ longer than the tibiæ; in the intermediate- and hind-legs the tibiæ and tarsi of equal length. Spurs very small. First joint of the tarsi twice the length of the second; second joint about $\frac{1}{4}$ longer than the third; third and fifth joints of equal length, about $\frac{1}{4}$ longer than the fourth. Wings pellucid, with an almost fuliginous tint; cupreous and violaceous reflections. Costal and two first longitudinal veins fuliginous. First longitudinal vein reaching the costa considerably before the base of the fork and somewhat before the tip of the posterior branch of the fourth longitudinal ; petiole paler than the fork, longer than the anterior branch ; branches, particularly the posterior one, little arcuated, making the fork rather angular at the base ; tips non-divergent. *fg* a little more than 3 times the length of *gh;* *kl* somewhat shorter than *lm.*

Hab.—Elizabeth Bay (Skuse). May.

2. Palpi yellow.

B. Cross-vein situated at the middle of the first longitudinal vein.

b. *Tip of the second longitudinal vein and tip of the posterior branch of the fork equally near the apex of the wing.*

123. Sciara luculenta, sp.n.

♀.—Length of antennæ......	0·085 inch	2·14 millimètres.
Expanse of wings.......	0·120 × 0·045	3·04 × 1·13
Size of body.............	0·120 × 0·020 ...	3·04 × 0·50

Antennæ slender, more than two-thirds the length of the body, black, with a short dense pubescence ; joints of the scapus light pitch-brown, with a few golden-yellow hairs longer than the pubescence of the flagellar joints ; flagellar joints sub-sessile, $2\frac{1}{3}$ to 4 times as long as broad, the terminal joint rather longer than the preceding. Head black, with a greenish reflection. Eyes contiguous above. Palpi pale yellow. Thorax ferruginous-ochraceous, levigate, with three longitudinal double rows of short golden-yellow hairs, the lateral ones also with a row of moderately long black hairs ; intermediate row reaching almost as far as the lateral ones, which extend nearly to the scutellum ; a few long black hairs, with some short golden yellow hairs, between the origin of the wings and the humeri ; scutellum with a few very long black hairs interspersed with a sparse short golden-yellow pubescence. Halteres pale brown with a few short hairs, stem honey-yellow. Abdomen deep brown, almost fuliginous on the dorsal segments, between the segments and underneath sordid ochraceous ; short moderately dense pubescence ; ovipositor long, ochraceous-brown, the lamellæ small, oval. Coxæ and femora pale yellow, the former with rather long brownish hairs in the upper side. Tibiæ and tarsi cinereous. In the fore- and intermediate-legs the tarsi a little longer than the tibiæ ; in the hind-legs the tibiæ and tarsi of about equal length. Spurs long, equal in length to the third tarsal joint. First joint of the tarsi about 3 times the length of the second ; second joint ¦ longer than the third and equal to the fourth and fifth together ; fourth joint same length as the fifth. Wings pellucid, with a pale yellowish-brown tint ; opaline reflections. Veins yellowish-brown. Fork of the third longitudinal vein and apical portion of the petiole fringed with short brown hairs. First longitudinal vein reaching the costa a little before the base of the fork, and considerably beyond the tip of the posterior branch of the fourth longitudinal ; petiole much paler than the fork, as long as the anterior branch ; branches running almost parallel to one another, scarcely divergent at the tips; posterior branch a little arcuated at the base; fourth longitudinal vein more distinct than the last. *fg* nearly twice the length of *gh* ; *kl* slightly longer than *lm*.

Hab.—Middle Harbour, near Sydney (Froggatt and Skuse). April.

c. *Tip of the posterior branch of the fork nearer the apex of the wing than the tip of the second longitudinal vein.*

124. Sciara fumipennis, sp.n.

♀.—Length of antennæ	0·050 inch	...	1·27 millimétres.
Expanse of wings	0·090 × 0·040	...	2·27 × 1·01
Size of body	0·085 × 0·015	...	2·14 × 0·38

Antennæ slender, more than half the length of the body, deep brown, with a dense pale brownish pubescence ; joints of the scapus deep brown, sparsely haired ; flagellar joints sub-sessile, 2½ to 3 times as long as wide, the terminal joint longer. Head black. Eyes contiguous above. Palpi yellow. Thorax reddish-brown, levigate, with three rather indistinct longitudinal double rows of very short brownish hairs extending nearly to the scutellum, not coalescent ; also a few short hairs between the origin of the wings and the humeri ; scutellum with some short brownish hairs and a few long brown setæ. Halteres light reddish-brown, with a few very short hairs, the base of the stem yellowish. Abdomen umber-brown on the dorsal segments, pale between the segments and underneath, rather densely clothed with a moderately long pubescence ; lamellæ of the ovipositor small, elongate. Coxæ honey-yellow, the femora more ochraceous ; tibiæ and base of the metatarsal joint dusky pitch-brown, the remainder black. In the fore-legs the tarsi nearly $\frac{1}{5}$ longer than the tibiæ ; in the intermediate-legs the tibiæ and tarsi of about equal length ; in the hind-legs the tibiæ a little longer than the tarsi. Spurs about the length of the third tarsal joint. First joint of the tarsi in the fore- and intermediate-legs rather more than twice the length of the second, in the hind-legs $2\frac{1}{2}$ times the length ; second joint $\frac{1}{4}$ longer than the third and somewhat longer than the fourth and fifth together ; third joint about $\frac{1}{3}$ longer than the fourth ; fourth and fifth joints of almost equal length. Wings pellucid, very pale smoky brown, with dusky brown veins ; brilliant margaritaceous reflections. First longitudinal vein reaching the costa a short distance before the base of the fork ; cross-vein distinct ; petiole very pale, shorter than the anterior branch of the fork ; both branches fringed with very short, stiff, brown hairs, running almost parallel to one another for the greater part of their length, and slightly divergent at the tips ; the posterior branch almost as much arcuated at the base as the anterior one. *tg* rather more than twice the length of *gh ; kl* equal to *lm.*

Hab.--Woronora (Masters and Skuse). October.

C. Cross-vein situated beyond the middle of the first longitudinal vein.

a. *Tip of the second longitudinal vein nearer the apex of the wing than the tip of the posterior branch of the fork.*

46

125. Sciara unica, sp.n.

♂.—Length of antennæ...... 0·060 inch   ...   1·54 millimètres.

Expanse of wings........ 0·095 × 0·040 ...   2·39 × 1·01

Size of body.............. 0·085 × 0·012 ...   2·14 × 0·30

Antennæ slender, about three-fourths the length of the body, black with a dense pale pubescence ; joints of the scapus black, very sparsely haired; flagellar joints sub-sessile (minute pedicels perceptible towards the tip) 2 to 3 times as long as broad, the terminal joint longer and more slender. Head black, levigate. Eyes contiguous above. Palpi yellowish. Thorax black, levigate, with three longitudinal single rows of very short golden-yellow hairs, running nearly parallel to one another ; also a few scattered short golden-yellow hairs on the lateral margins before the origin of the wings, and on the scutellum. Halteres cinereous, the base of the stem ochraceous ; club sparsely covered with very short hairs. Abdomen slender, black, sparingly clothed with some moderately long pale hairs ; forceps rather elongate, not so wide as the terminal abdominal segment. Coxæ and femora ochraceous; tibiæ and first joint of the tarsi ochraceous-brown, the remaining joints black. In all the legs the tarsi a little longer than the tibiæ, about ¼ longer in the fore-legs. Spurs pale yellow, shorter than the fourth tarsal joint. First joint of the tarsi nearly 2¼ times the length of the second ; second joint ⅓ longer than the third, and as long as the fourth and fifth together ; third joint ½ longer than the fourth ; fifth joint a little longer than the fourth. Wings pellucid, with a somewhat fuliginous tint, costal and two first longitudinal veins fuliginous ; margaritaceous reflections. First longitudinal vein reaching the costa a short distance before the base of the fork, and opposite the tip of the posterior branch of the fourth longitudinal ; petiole very indistinct, shorter than the posterior branch of the fork ; branches almost equally arcuated at the base, running nearly parallel for the greater part of their length, divergent at the tips ; tip of the second longitudinal vein somewhat nearer the apex of the wing than the tip of the posterior branch. *fg* twice the length of *gh*; *kl* almost as long as *lm*.

Hab.—Gosford (Skuse). February.

b. *Tip of the second longitudinal vein and tip of the posterior branch of the fork equally near the apex of the wing.*

126. Sciara Winnertzi, sp.n.

♂.—Length of antennæ...... 0·110 inch ... 2·79 millimètres.

Expanse of wings........ 0·100 × 0·040 ... 2·54 × 1·01

Size of body.............. 0·110 × 0·015 ... 2·79 × 0·38

Antennæ slender, black, with a dense pale pubescence; joints of the scapus obscure pitch-brown, with very few hairs; flagellar joints sub-sessile, three to five times as long as broad, the terminal joint considerably longer; pedicels very short. Head black. Eyes contiguous above. Palpi honey-yellow. Thorax obscure pitch-brown, levigate with three longitudinal double rows of pale brownish hairs, the intermediate one running about two-thirds of the distance to the scutellum, the lateral ones a little convergent, reaching almost to the scutellum; also some setaceous hairs between the origin of the wings and the humeri; humeri tipped with ochraceous-brown; scutellum pitch-brown, with a few setaceous hairs. Halteres light pitch-brown, with a few short hairs, the stalk honey-yellow. Abdomen slender, dorsal segments deep brown, between the segments and underneath honey-yellow; densely clothed with a moderately long pale brownish pubescence; forceps pitch-brown, wider than the last two segments, densely pubescent. Coxæ and femora honey-yellow, the former with a somewhat reddish tinge, and long hairs on the front. Tibiæ pitch-brown. Tarsi almost black, on account of their dense pubescence; the metatarsal joint brownish at the base. In the fore-legs the tarsi somewhat longer than the tibiæ; in the intermediate-legs the tibiæ and tarsi of about equal

length ; in the hind-legs the tibiæ somewhat longer than the
tarsi. Spurs almost as long as the third tarsal joint. First joint
of the tarsi about 3 times the length of the second ; second joint
¼ longer than the third and almost as long the third and fourth
together ; third joint ⅓ longer than the fourth and almost twice
the length of the fifth. Wings pellucid, with a very pale brownish-
yellow tint; veins pale brown; brilliant margaritaceous reflections
when viewed at a certain obliquity. First longitudinal vein
reaching the costa a short distance before the base of the fork,
and opposite to the tip of the posterior branch of the fourth
longitudinal vein ; cross-vein short and indistinct, just beyond the
middle of the first longitudinal vein ; petiole very pale and indistinct, not quite as long as the anterior branch of the fork ;
branches running parallel for the greater part of their length and
slightly divergent at the tips ; posterior branch very little arcuated
at the base. *fy* about twice the length of *gh* ; *kl* a little shorter
than *lm*.

Hab.—Glenbrook (Masters). November.

127. SCIARA MONTIVAGA.

♂.—Length of antennæ...... 0·070 inch   ... 1·77 millimètres.
  Expanse of wings......... 0·090 × 0·035 ... 2·27 × 0·88
  Size of body............. 0·080 × 0·012 ... 2·02 × 0·30

Antennæ slender, not quite the length of the body, deep brown,
with a dense pale yellowish pubescence ; joints of the scapus deep
brown, with very little pubescence ; flagellar joints sub-sessile, 2 to
4½ times as long as broad, the terminal joint longer. Head black.
Eyes contiguous above. Palpi yellow. Thorax black, levigate,
with three longitudinal double rows of short brownish-yellow
hairs, the intermediate row extending almost as far as the lateral
ones, these latter almost reaching the scutellum ; also some similar
hairs along the lateral margins and on the scutellum, interspersed
with a few long brown setæ. Halteres light brown, the base of
the stalk yellowish, sparsely covered with very short hairs.

Abdomen about as wide as the thorax, black or very deep brown
on the dorsal segments, pale underneath, densely clothed with a
short pubescence; forceps not so wide as the thorax, densely
pubescent. Coxæ and femora honey-yellow, tibiæ and tarsi pale
pitch-brown. In the fore-legs the tarsi $\frac{1}{4}$ longer than the tibiæ;
in the intermediate-legs the tarsi somewhat longer than the tibiæ;
in the hind-legs the tibiæ a little longer than the tarsi. Spurs
honey-yellow, about as long as the fourth tarsal joint. First joint
of the tarsi twice the length of the second; second joint $\frac{1}{4}$ longer
than the third and longer than the fourth and fifth together;
third joint $\frac{1}{3}$ longer than the fourth; fourth joint rather longer
than the fifth. Wings hyaline, the costal and two first longitudinal veins brown, the rest very pale; brilliant yellow-green
and roseous reflections. First longitudinal vein reaching the
costa a short distance before the base of the fork; cross-vein
rather pale; petiole almost invisible, longer than the anterior
branch of the fork; both branches very indistinct at the base,
running almost parallel for the greater part of their length, a
little divergent at the tips; the posterior branch only very slightly
arcuated at the base. *fg* twice the length of *gh*; *kl* rather shorter
than *lm*.

Hab.—Glenbrook (Masters); Berowra and Knapsack Gully
(Masters and Skuse). August to November.

128. SCIARA ORNATULA, sp.n.

♂.—Length of antennæ...... 0·055 inch   ... 1·89 millimètres.
Expanse of wings........ 0·065 × 0·025 .. 1·66 × 0·62
Size of body .... 0·065 × 0·010 ... 1·66 × 0·25

Antennæ slender, not quite the length of the body, deep brown,
with a dense pale yellowish pubescence; joints of the scapus deep
brown, very sparsely pubescent; flagellar joints $2\frac{1}{2}$ to $4\frac{1}{2}$ times as
long as broad, with very short pale pedicels, the terminal joints
very slender. Head black. Eyes contiguous above. Palpi
yellow. Thorax light ferruginous-brown, levigate, with three

longitudinal double rows of very short brownish-yellow hairs, the intermediate row extending a little beyond the middle of the thorax, the lateral ones not quite reaching the scutellum; a very few long setæ on the lateral margins above the origin of the wings, also two long setæ on the scutellum, with a sparse sprinkling of short brownish-yellow hairs. Halteres obscure umber - brown, the root of the stem yellowish, club sparsely covered with very short hairs. Abdomen dusky umber-brown, whitish between the segments, somewhat sparsely clothed with a short pubescence; forceps light ferruginous-brown, rather small, considerably narrower than the abdomen, densely covered with a minute pubescence. Coxæ and femora honey-yellow, the tibiæ and tarsi darker on account of their minute dense pubescence. In the fore-legs the tarsi about $\frac{1}{5}$ longer than the tibiæ ; in the intermediate-legs the tarsi somewhat longer than the tibiæ ; and in the hind-legs the tibiæ a very little longer than the tarsi. Spurs honey-yellow, shorter than the fourth tarsal joint. First joint of the tarsi in the two first pairs of legs about twice the length of the second, in the hind-legs rather more than twice its length ; second joint about $\frac{1}{4}$ longer than the third and equal in length to the fourth and fifth together ; third joint about $\frac{1}{3}$ longer than the fourth ; fourth and fifth joints of equal length. Wings hyaline, the costal and two first longitudinal veins brownish-yellow, the rest pale ; purpureous reflections. First longitudinal vein reaching the costa some distance before the base of the fork ; cross-vein rather pale ; petiole a little paler than the fork, longer than the anterior branch ; branches running almost parallel for the greater part of their length, tips a little divergent. *fy* about $3\frac{1}{2}$ times the length of *yh* ; *kl* shorter than *lm*.

Hab.—Sydney (Skuse).   September.

129. Sciara Amabilis, sp.n.

♂.—Length of antennæ...... 0·045 inch   ...   1·13 millimètres.
Expanse of wings........ 0·085 × 0·035 ...   2·14 × 0·88
Size of body.............. 0·080 × 0·012 ...   2·02 × 0·30

Antennæ slender, rather more than half the length of the body, black, densely covered with a very short pale pubescence ; joints of the scapus black, sparsely haired ; flagellar joints sub-sessile, 2 to 3 times as long as broad, becoming very slender towards the tip. Head black. Eyes contiguous above. Palpi yellow. Thorax black, sub-nitidous, with three longitudinal double rows of short yellowish-brown hairs, the intermediate row extending for about three-fourths of the distance to the scutellum, the lateral ones almost reaching the scutellum, not coalescent; some short yellowish-brown hairs, interspersed with long brown setæ, on the lateral margins between the origin of the wings and the humeri, and on the scutellum. Halteres dusky-brown, with a sprinkling of very short hairs, the base of the stem brownish-ochraceous. Abdomen as wide as the thorax, deep brown, appearing almost black, densely clothed with a moderately long brownish pubescence; forceps deep brown, broader than the terminal abdominal segment. Coxæ ferruginous-ochraceous ; femora, tibiæ and tarsi dusky pitch-brown, the last three joints of the tarsi nearly black. In the fore-legs the tarsi nearly ½ longer than the tibiæ; in the intermediate-legs the tarsi about ⅓ longer than the tibiæ ; in the hind-legs the tibiæ and tarsi of equal length. Spurs shorter than the last tarsal joint. First joint of the tarsi twice the length of the second ; second joint ¼ longer than the third and equal to the fourth and fifth together; third joint about ½ longer than the fourth ; fourth joint somewhat longer than the fifth. Wings almost hyaline, but having a faint greyish tint, the costal and two first longitudinal veins brown, the rest of the veins greyish ; violaceous and purpureous reflections. First longitudinal vein reaching the costa a short distance before the base of the fork ; cross-vein somewhat indistinct ; petiole hardly visible, a very little longer than the anterior branch of the fork ; both branches distinct, the posterior branch less arcuated at the base than the anterior one, running almost parallel to one another for the greater part of their length, the tips considerably divergent ; fourth longitudinal vein distinct. *fg* about ½ longer than *gh* ; *kl* somewhat shorter than *lm*.

Hab.—Sydney (Masters and Skuse). September.

130. Sciara lucidipennis, sp.n.

♀.—Length of antennæ...... 0·030 inch   ...   0·76 millimètre.
    Expanse of wings....... 0·070 × 0 025 ...   1·77 × 0·62
    Size of body .... 0·070 × 0·015 ...   1·77 × 0·38

Antennæ deep brown, with a minute pale pubescence ; slender, not half the length of the body; joints of the scapus rather paler brown than those of the flagellum, with very little pubescence ; flagellar joints sub-sessile, 2 to 2½ times as long as broad, densely haired. Head black, levigate. Eyes contiguous above. Palpi yellow. Thorax obscure reddish-black, with three longitudinal double rows of minute yellowish-brown hairs from the collare to the scutellum, also a few long hairs on the lateral margins between the origin of the wings and the humeri, and on the scutellum ; pleuræ obscure reddish-brown. Halteres obscure olivaceous, with a few very minute hairs on the apex, stem yellow. Abdomen broader than the thorax, olivaceous, paler between the segment and on the underside, rather sparsely clothed with a minute pubescence; lamellæ of the ovipositor minute, elliptical, rather densely pubescent. Legs greyish-ochraceous, the front of the femora, and the tibiæ and tarsi considerably darker than the rest. Coxæ with tolerably long hairs on the front. Femora, tibiæ and tarsi with a minute pubescence, much less dense on the femora. In the fore- and intermediate-legs the tarsi a little longer than the tibiæ ; in the hind-legs the tibiæ and tarsi of about equal length. First joint of the tarsi shorter than the four following combined, and 2¼ times the length of the second ; second joint about ¦ longer than the third and shorter than the fourth and fifth together ; fourth joint somewhat shorter than the fifth. Wings almost hyaline, with a very pale brownish tint; the costal and two first longitudinal veins obscure yellowish-brown ; tip and posterior margin with brilliant cupreous reflections, the anterior portion violaceous and bright æneous. First longitudinal vein reaching the costa considerably before the base of the fork and opposite to the tip of the posterior branch of the fourth longitudinal vein ; petiole paler than the fork, and longer than the anterior branch ; posterior branch very little arcuated at the base ; anterior branch more divergent at the tip than the posterior branch. *fg* about $2\frac{1}{4}$ times the length of *gh ; kl* shorter than *lm.*

Hab.—Elizabeth Bay (Skuse). April.

c. *Tip of the posterior branch of the fork nearer the apex of the wing than the tip of the second longitudinal vein.*

131. SCIARA NUBICULA, sp.n.

♀.—Length of antennæ...... 0·060 inch   ... 1·54 millimètres.
Expanse of wings. ...... 0·095 × 0·035 .. 2·39 × 0·88
Size of body............... 0·090 × 0·015 ... 2·27 × 0·38

Antennæ very slender, two-thirds the length of the body, deep brown, densely pubescent ; joints of the scapus pitch-brown, sparingly haired ; flagellar joints sub-sessile, 3 to $3\frac{1}{2}$ times as long as broad. Head black. Eyes contiguous above. Palpi yellow. Thorax deep brown, sub-nitidous, with three longitudinal single rows of brown hairs, double just before the collare, also a few very long setæ just before the origin of the wings and on the scutellum. Halteres dusky brown, with a very short sparse pubescence, the base of the stem yellowish. Abdomen umber-brown on the dorsal segments, whitish between the segments and underneath, densely clothed with a short brown pubescence ; lamellæ of the ovipositor very small, oval. Coxæ and femora ochraceous ; tibiæ and tarsi ochraceous-brown, the last joints of the tarsi dusky. In the fore-legs the tarsi nearly $\frac{1}{4}$ longer than the tibiæ ; in the intermediate-legs the tarsi about $\frac{1}{5}$ longer than the tibiæ ; in the hind-legs of equal length. Spurs honey-yellow, about the length of the fourth tarsal joint. First joint of the tarsi in the fore- and intermediate-legs twice the length of the second, in the hind-legs $2\frac{1}{2}$ times the length ; second joint $\frac{1}{5}$ longer than the third and equal to the fourth and fifth together ; third joint $\frac{1}{4}$ longer than the fourth ; fourth joint about

⅕ longer than the fifth. Wings pellucid with a greyish tint, the costal and two first longitudinal veins umber-brown; brilliant opaline reflections. First longitudinal vein reaching the costa a short distance before the base of the fork; cross-vein distinct; third longitudinal vein very pale; petiole scarcely visible, longer than the anterior branch; fork narrow, the branches almost equally arcuated at the base, running parallel to one another for the greater part of their length, a little divergent at the tips. *fg* twice the length of *gh*; *kl* somewhat shorter than *lm*.

Hab.—Middle Harbour (Froggatt and Skuse). April.

132. SCIARA SPECTABILIS, sp.n.

♂.—Length of antennæ... .. 0·070 inch   ... 1·77 millimètres.
Expanse of wings... .... 0·080 × 0·030 ... 2·02 × 0·76
Size of body.......... .... 0·080 × 0·012 ... 2·02 × 0·30

♀.—Length of antennæ...... 0·045 inch   ... 1·13 millimètres.
Expanse of wings........ 0·095 × 0·040 ... 2·39 × 1·01
Size of body .... 0·095 × 0·015 ... 2·39 × 0·38

Antennæ slender, nearly the length of the body, deep brown, with a dense brownish-yellow pubescence; joints of the scapus sparingly haired, the first joint ochraceous, the second deep brown; flagellar joints sub-sessile, 2½ to 4 times as long as broad, the terminal joints very slender. Head black. Eyes contiguous above. Palpi yellow. Thorax pitch-brown, levigate, with three darker narrow longitudinal stripes, each with a single row of deep brown hairs, the intermediate one stopping a short distance from the scutellum, the lateral ones not coalescent, reaching the scutellum; the lateral margins and scutellum setose. Halteres dusky-brown, the base of the stem yellow, club sparingly covered with a short pubescence. Abdomen brown on the dorsal segments, whitish underneath and between the segments; densely pubescent; forceps brown, a little wider than the terminal segment. Coxæ and femora honey-yellow; tibiæ and tarsi almost cinereous. In the fore-legs the tarsi ⅓ longer than the tibiæ; in the intermediate-legs the tarsi almost imperceptibly longer than the tibiæ ; in the hind-legs the tibiæ nearly ⅕ longer than the tarsi. Spurs honey-yellow, as long as the last tarsal joint. First joint of the tarsi about 3 times the length of the second ; second joint ⅙ longer than the third ; third joint ½ longer than the fourth ; fifth joint somewhat longer than the fourth. Wings pellucid, with a pale yellowish tint, the costal and two first longitudinal veins yellowish-brown ; brilliant margaritaceous reflections. First longitudinal vein reaching the costa some distance before the base of the fork and somewhat before the tip of the posterior branch of the fourth longitudinal ; petiole almost invisible, considerably longer than either branch of the fork ; branches running almost parallel for the greater part of their length, slightly divergent at the tips, the posterior branch very little arcuated at the base ; the petiole just before the fork and both branches ciliate, in some specimens more so than in others. *fy* rather more than 2½ times the length of *gh* ; *kl* somewhat shorter than *lm*.

♀.—Antennæ a little longer than the head and thorax together; joints of the scapus ochraceous ; flagellar joints 2 to 3 times as long as broad. Thorax brown-ochraceous, with no darker stripes but two distinct rows of deep brown hairs, and a very indistinct intermediate one, extending a little beyond the middle of the thorax. Abdomen of a lighter brown on the dorsal segments than in the ♂ ; ovipositor pale brown, the lamellæ small, elongate. Tibial spurs longer than the fourth and fifth tarsal joints. Petiole almost invisible, about equal in length to the anterior branch of the fork. Cross-vein not so much beyond the middle of the first longitudinal vein as in the ♂. *fy* three times the length of *gh* ; *kl* almost as long as *lm*.

Hab.—Sydney and Berowra (Masters and Skuse). November to January.

133. SCIARA IGNOBILIS, sp.n.

♂.—Length of antennæ	0·045 inch	...	1·13 millimètres.
Expanse of wings	0·070 × 0 030	...	1·77 × 0·76
Size of body	0·065 × 0·012	...	1·66 × 0·30

Antennæ slender, nearly three-fourths the length of the body, black, with a dense short yellowish pubescence ; joints of the scapus black, very sparsely haired ; flagellar joints visibly pedicelled, the pedicels ⅕ the length of the joints; joints twice as long as broad, the terminal one a little longer. Head black. Eyes contiguous above. Palpi yellowish. Thorax black, levigate, with three longitudinal double rows of yellowish hairs, the intermediate row scarcely perceptible, apparently terminating before the middle of the thorax, and the lateral ones extending nearly to the scutellum, the hairs sparse, long ; also some long yellowish setæ on the lateral borders, and on the scutellum. Halteres dusky brown, with a few short hairs, the stem yellowish. Abdomen as broad as the thorax, obscure olivaceous, pale between the segments and underneath ; rather sparsely haired ; forceps rather wider than the terminal segment, densely pubescent. Coxæ and femora yellowish-brown ; tibiæ and tarsi dusky brown, the last four tarsal joints black. In the fore-legs the tarsi longer than the tibiæ by the last joint ; in the intermediate-legs the tibiæ and tarsi almost of equal length, the tibiæ being somewhat longer ; in the hind-legs the tibiæ a very little longer than the tarsi. Spurs shorter than the fourth tarsal joint. First joint of the tarsi 2½ times the length of the second; second joint ⅓ longer than the third ; third joint about ¼ longer than the fourth and somewhat longer than the fifth. Wirgs almost hyaline, but having a yellowish tint, the costal and first two longitudinal veins umber-brown ; rather weak opaline reflections. First longitudinal vein reaching the costa some distance before the base of the fork ; cross-vein somewhat indistinct ; petiole much paler than the fork and shorter than the anterior branch ; branches almost imperceptibly divergent for the whole of their length, not at the tips ; the posterior branch very little arcuated at the base ; branches of the fourth longitudinal vein more distinct than the last. *fg* nearly twice *gh*; *kl* somewhat longer than *lm*.

Hab.—Berowra (Masters and Skuse). August.

134. Sciara infrequens, sp.n.

♂.—Length of antennæ...... 0·040 inch   ...   1·01 millimètres.
   Expanse of wings....... 0·060 × 0·025 ... 1·54 × 0·62
   Size of body.............. 0·055 × 0·010 ... 1·39 × 0·25

Antennæ slender, rather more than two-thirds the length of the body, brown, with a dense brownish-yellow pubescence; joints of the scapus with scarcely any pubescence; flagellar joints sub-sessile, 3 to 4 times as long as broad, the joints very slender towards the tip. Head black. Eyes contiguous above. Palpi yellow. Thorax reddish-brown, levigate, with three longitudinal single rows of brownish hairs from the collare to the scutellum, not coalescent posteriorly; also a few brown setæ between the origin of the wings and the humeri and on the scutellum. Halteres dusky-brown, sprinkled with a few very short hairs, the stem yellowish at the base. Abdomen slender, black or very deep brown on the dorsal segments, pale between the segments and underneath, densely clothed with a moderately long pubescence; forceps somewhat elongate, narrower than the terminal segment, ochraceous-brown, densely pubescent. Coxæ and femora honey-yellow; tibiæ and tarsi ochraceous. In the fore-legs the tarsi somewhat longer than the tibiæ; in the intermediate-legs the tibiæ somewhat longer than the tarsi; in the hind-legs the tibiæ ⅙ longer than the tarsi. Spurs as long as the fourth tarsal joint. First joint of the tarsi about 2½ times the length of the second; second joint ¼ longer than the third, and equal to the fourth and fifth together; third joint almost ⅓ longer than the fourth; fourth joint somewhat shorter than the fifth. Wings hyaline, the costal and two first longitudinal veins umber-brown; smaragdine and golden reflections. First longitudinal vein reaching the costa considerably before the base of the fork; cross-vein rather indistinct; petiole almost invisible, somewhat longer than the anterior branch of the fork; branches moderately arcuated at the base, the posterior less than the anterior, running almost parallel to one another for the greater part of their length, slightly divergent at the tips. *fg* nearly 3 times the length of *gh*; *kl* somewhat shorter than *lm*.

Hab.—Elizabeth Bay (Skuse). January.

B. Halteres yellow or whitish.

        1. Palpi black or brown.

B. Cross-vein situated at the middle of the first longitudinal vein.

  c. *Tip of the posterior branch of the fork nearer the apex of the wing than the tip of the second longitudinal vein.*

135. Sciara notata, sp.n.

♂.—Length of antennæ......	0·095 inch	...	2·39 millimètres,
Expanse of wings.........	0·105 × 0·040	...	2·67 × 1·01
Size of body	0·100 × 0·017	...	2·54 × 0·42

Antennæ slender, almost the length of the body, black, with a short, dense brownish pubescence; first joint of the scapus black, second pitch-brown, sparsely haired; flagellar joints with very minute pedicels, the joints 3 to 6 times as long as broad, those towards the tip being very slender. Head black. Eyes contiguous above. Palpi brown. Thorax black, levigate, with three longitudinal rows of golden yellow hairs, the intermediate row single, not very distinct, extending a little beyond the middle of the thorax, the lateral ones double, the hairs much longer than those of the intermediate row, and very long just before the scutellum; also some golden-yellow pubescence and brown setæ on the lateral borders and on the scutellum. Halteres honey-yellow. Abdomen not so wide as the thorax, black in the dorsal segments, deep brown underneath, densely pubescent; forceps short, robust, rather wider than the terminal segment. Coxæ and femora bright pitch-brown; tibiæ and tarsi dusky brown, the tips of the tarsi fuliginous. In the fore-legs the tarsi about ⅓ longer than the tibiæ ; in the intermediate-legs the tarsi a very little longer than the tibiæ ; in the hind-leg the tibiæ and tarsi very long, of equal length. Spurs as long as the fourth tarsal joint. First tarsal joint in the fore- and intermediate-legs about twice the length of the second, in the hind-legs $2\frac{2}{3}$ times the length ; second joint about $\frac{1}{5}$ longer than the third, and rather longer than the fourth and fifth joints together ; third joint about $\frac{1}{3}$ longer than the fourth ; fourth joint a little longer than the fifth. Wings pellucid with a pale smoky tint, the costal and two longitudinal veins umber-brown ; pale opaline reflections. First longitudinal vein reaching the costa a short distance before the base of the fork ; cross-vein somewhat indistinct ; petiole very pale, about as long as the anterior branch of the fork ; branches running almost parallel to one another, a little divergent at the tips ; the posterior branch very little arcuated at the base, its tip very little nearer the apex of the wing than the tip of the second longitudinal vein ; fourth longitudinal vein with its branches well-defined. *fg* about $2\frac{1}{3}$ times the length of *gh* ; *kl* a little shorter than *lm*.

Hab.—Glenbrook, Blue Mountains (Masters). November.

2. Palpi yellow.

C. Cross-vein situated beyond the middle of the first longitudinal vein.

b. *Tip of the second longitudinal vein and tip of the posterior branch of the fork equally near the apex of the wing.*

136. SCIARA PICTIPES, sp.n.

♂.—			
Length of antennæ	0·075 inch	...	1·89 millimètres.
Expanse of wings	0·090 × 0·035	...	2·27 × 0·88
Size of body	0·080 × 0.015	...	2·02 × 0·38

Antennæ slender, particularly towards the tip, nearly the length of the body, black, with a very dense greyish-brown pubescence ;

joints of the scapus very sparsely haired ; flagellar joints sub-
sessile, 2¼ to 4 times as long as broad, the terminal joint considerably longer. Head black. Eyes contiguous above. Palpi
yellow. Thorax black, levigate, with three longitudinal double
rows of minute yellowish pubescence, the intermediate row
indistinct, reaching a little beyond the middle of the
thorax, the lateral ones very distinct, extending almost to the
scutellum, not coalescent ; some short yellowish pubescence and
long setæ between the origin of the wings and the humeri,
also on the scutellum. Halteres pale yellow, with very little
visible pubescence. Abdomen nearly as wide as the thorax,
uniformly black, moderately clothed with a short yellowish
pubescence ; forceps short and robust, as wide as the terminal
abdominal segment. Coxæ and femora ochraceous ; tibiæ and
tarsi brownish-ochraceous, the tips of the tarsi dusky. In the
fore-legs the tarsi nearly ¼ longer than the tibiæ ; in the intermediate-legs the tarsi somewhat longer than the tibiæ ; in the
hind-legs the tibiæ about ¹⁄₁₀ longer than the tarsi. Spurs as long
as the last tarsal joint. First joint of the tarsi in the fore- and
intermediate-legs about twice the length of the second, in the
hind-legs 2⅔ times the length ; second joint about ⅙ longer than
the third and somewhat longer than the fourth and fifth together ;
third joint ⅓ longer than the fourth ; fourth joint ¼ longer than
the fifth. Wings pellucid, with a very pale smoky appearance,
the costal and two first longitudinal veins deep brown ; bright
margaritaceous reflections. First longitudinal vein reaching the
costa a short distance before the base of the fork ; cross-vein distinct, situated a little beyond the middle of the first longitudinal ;
petiole very pale, shorter than the anterior branch of the fork ;
branches running almost parallel to one another for the greater
part of their length, divergent at the tips ; the posterior branch
very little arcuated at the base. *fg* nearly twice the length of
gh ; *kl* somewhat shorter than *lm*.

Hab.—In the neighbourhood of Narrabeen Lagoon (Skuse).
October.

Genus 2. TRICHOSIA, Winnertz.

Trichosia, Winnertz, Beitr. z. Mon. d. Sciarinen, 1867.

Characters the same as in *Sciara*, with the difference that the surface of the wings is distinctly hairy.

Only a very few species of this distinct genus have yet been described, but they come from widely different parts of the world, so that this genus is no doubt represented in most countries. The following is the first described from Australia; others have been recorded from Europe and North America :—

137. TRICHOSIA MASTERSI, sp.n.

♂.—Length of antennæ...... 0·088 inch   ... 2·14 millimètres.

Expanse of wings........ 0·090 × 0·035 ... 2·27 × 0·88

Size of body.............. 0·085 × 0·020 ... 2·14 × 0·50

♀.—Length of antennæ...... 0·047 inch   ... 1·23 millimètres.

Expanse of wings........ 0·100 × 0·040 ... 2.54 × 1.01

Size of body.............. 0·110 × 0·030 ... 2·79 × 0·76

♂.—Antennæ pitch-brown, with a minute yellowish pubescence; slender, the length of the body; basal joints deep brown, sparsely pubescent; flagellar joints sub-sessile, 2½ to 3 times as long as broad. Head black. Eyes contiguous above. Palpi black or deep brown. Thorax black, levigate, with three longitudinal double rows of yellowish hairs, the intermediate one indistinct; setaceous hairs on the lateral margins and scutellum. Halteres almost naked, smoky-yellow, with the base of the stem a brighter yellow; club large, pyriform. Abdomen black, with a pale pubescence, narrower than the thorax, almost cylindrical, gradually dilated towards the middle; forceps black, densely covered with a minute pubescence, considerably broader than the terminal segment of the abdomen. Legs yellowish-brown, the under-sides of all the femora considerably brighter than the other portions of the legs; tarsal joints almost pitch-brown on account of their minute dense pubescence. Fore femora shorter

47

than the intermediate ones, and the latter shorter than the hind femora. In the fore-legs the tarsi longer than the tibiæ ; in the intermediate-legs both of about equal length ; in the hind-legs the tibiæ somewhat longer than the tarsi. Spurs bright honey-yellow. First joint of the tarsi rather more than twice the length of the second ; second joint almost one-half longer than the third ; fourth and fifth joints of nearly equal length. Wings with a brownish-grey tint ; densely covered with micro-scopic pubescence intermixed with some irregularly dispersed and considerably longer, bent hairs ; rather brilliant pale green and rosy reflections. Veins brownish-grey. First longitudinal vein joining the costa before the base of the fork ; cross-vein situated beyond the middle of the first longitudinal vein ; tip of the second longitudinal vein and the tip of the posterior branch of the fork equally near the apex of the wing ; petiole very pale and indistinct, longer than the anterior branch of the fork ; branches of the fork slightly divergent at the tip ; fourth longitudinal vein pale but distinct ; anterior branch of the fourth longitudinal vein very little arcuated, posterior branch running rather close to the anterior branch for half its length, then turning somewhat abruptly to the posterior margin. fg almost twice the length of gh ; kl a very little shorter than lm. Rudimentary fifth longitudinal vein rather close to the fourth longitudinal before branching, running very close to the posterior branch, and disappearing at about $\frac{2}{3}$ of its length.

♀.—Antennæ longer than the head and thorax together ; flagellar joints sub-sessile, 2 to $2\frac{1}{2}$ times as long as broad. Thorax more densely pubescent than that of the ♂, the lateral longi-tudinal rows of hairs apparently treble. Halteres dusky. Legs dusky-brown with a yellowish tint, having a coarser pubes-cence than in the ♂. Abdomen pale umber-brown, with dusky reflections. Wings very densely covered with a somewhat inter-woven pubescence, making the third longitudinal vein very indis-tinctly visible.

Hab.—Como (Masters and Skuse). September.

EXPLANATION OF PLATE.

PLATE XI.

Fig. 1. Alar-venation of *Sciara Macleayi.*

Fig. 2. ,, ,, *sedula.*

Fig. 3. ,, ,, *æmula.*

Fig. 4. ,, *Trichosia Mastersi.*

Fig. 5. Diagram illustrating the terminology for the veins and cells as applied to the Sciaridæ.

[*The dextral column gives the German equivalents employed by Winnertz, (Beit. zu einer Mon. der Sciarinen, 1867)*].

Veins.	*Adern.*
Costa *(v. costalis).* a, e, g.	Randader.
Transverse shoulder-vein *(v. transversa humeralis).* b.	Hülfsader.
Auxiliary *(v. auxiliaris).* c.	
1st longitudinal *(v. long. 1ma).* a, e.	Unterrandader.
Cross-vein *(v. transvers marginalis).* d.	Querader.
2nd longitudinal *(v. long. 2da).* a, f.	Mittelader + Ellbogenader.
Anterior branch *(v. long. 2da ramus anterior).* s.	Brachialader.
3rd longitudinal *(v. long. 3a).* n, p, k.	Mittlere Scheibenader.
Anterior branch *(v.long.3a ramus anterior).* h.	Obere Scheibenader.
4th longitudinal *(v. long. 4a).* a, m.	Hinterader.
Anterior branch *(v. long. 4a ramus anterior).* l.	Untere Scheibenader.
5th longitudinal *(v. long. 5a).* n. *(Rudimentary).*	Achselader.

Cells.		*Zellen.*
Sub-costal *(c. subcostalis).* A.		Randzelle.
Inner marginal *(c. marginalis interior).* B.		Schulterzelle.
Marginal *(c. marginalis).* C.		Vordere Cubitalzelle.
1st sub-marginal *(c. submarginalis 1ma).* D.		Hintere Cubitalzelle.
2nd sub-marginal *(c. submarginalis 2da).* E.		Obere Scheibenzelle.
3rd sub-marginal *(c. submarginalis 3a).* F.		Mittlere Scheibenzelle.
1st posterior *(c. posterior 1ma).* G.		Untere Scheibenzelle.
2nd posterior *(c. posterior 2da).* H.		Hinterzelle.
Axillary *(c. axillaris).* I.		Achselzelle.

www.ingramcontent.com/pod-product-compliance
Lightning Source LLC
Chambersburg PA
CBHW022002190326
41519CB00010B/1367